"Vision is the art of seeing things invisible."

Jonathan Swift

BriefBook®

Biotechnology, Microbes and the Environment

by

Steven C. Witt

This *BriefBook* is for
my mother,
my bride,
freda and klyde.

Contents

Appendix

Some Story Ideas in this *BriefBook*

As you read this ***BriefBook****, different things — a compelling fact, a telling quote — will ignite sparks in your mind that make you want to know more. Here are some examples:*

Brief Look at this BriefBook

"My daughter Alice is a student in High School. One of the prescribed courses is General Science. The section on Bacteria left her with a vague impression of a world teeming with deadly germs awaiting an opportunity to infect mankind. It seems probable that this malignant conception of bacteria is very generally held. It is deplorable that the introduction to Science, the key to Nature's wonders, should be darkened by a vision of an unmitigated, hidden force for evil, standing squarely in opposition to man. In reality, civilization owes much to the microbe."

Arthur Isaac Kendall, *Civilization and the Microbe,* 1923

This *BriefBook* is about things small — very, very small. It is about microorganisms, commonly called microbes. How small is microbial small? Well, 1,000 bacteria, one kind of microorganism, strung end-to-end like pearls on a necklace would barely stretch across the period at the end of this sentence.

Like Kendall's daughter Alice, you probably have a negative impression of microbes. After all, microbes do make misery —

"Whatever it is, it's very, very little."

Drawing by Stevenson; © 1985 The New Yorker Magazine, Inc.

flus and colds, mumps and measles, even certain cancers.Nonetheless, as Kendall realized, the world of microspaces beyond our vision where periods look like good-sized buildings, is truly a natural wonderland, and the overwhelming majority of the 200,000-plus known species of microbes do positive, not harmful, things.

But most people have never been formally exposed to microbiology. So they don't really know the first thing about microbes or the microspaces they live in.

The Microspace Shuttle

But they will find out. They won't be able to avoid it. And biotechnology is a big reason why. You see, back in 1973 when two California scientists spliced a toad gene into a microbe and started the "biotech revolution," they sent science on a fantastic voyage into microspace. And now biotech is propelling scientists in laboratories around the globe on millions of missions into the minute world of microbes. Biotechnology is the space shuttle not of the cosmos, but of the microcosmos.

Because of biotechnology, the heretofore uncrackable black boxes that make microbiology a science defined mostly by darkness are now being pried open, and secret after secret is spilling out into the full light of scientific examination and public scrutiny. The pace of this implosion into the minuscule will continue to increase into the 21st century.

Microbes Outside, Debates Inside

Biotech is also enabling scientists to change and manipulate microbes in ways that seemed the stuff of science fiction just a few years ago. The developers of these new microbes are eager to see whether they work — as products that kill pests, clean up oil spills and toxic chemicals, fertilize crops and much more. And that means releasing genetically altered microbes *outside* in the environment. Several different kinds of microbes — including some that Mother Nature could not create — already have been released into fields from California to Maryland, the United Kingdom to Australia.

These outdoor releases have sparked vigorous debates around the globe, causing major rifts in the scientific community, anxiety in environmental circles, nightmares for government regulators and sweet dreams of big payoffs for business executives.

All before the vast majority of the world knows even the basic facts about microorganisms.

Meet the simplest kind of microbe and the star of this BriefBook, the basic bacterium.

Thus, this *BriefBook.* Its goal is to inform. It is based on two years of original research conducted by the Center for Science Information, highlighted by more than 300 hours of in-depth interviews with experts from around the world, listed in "Expert Sources" on page 207. This *BriefBook* will open up the mysterious world of microbes, yielding glimpses of the wonder in the unseeable microspaces all around us. It is your ticket to preview one of the truly fantastic voyages of the 21st century: the journey into microspace made possible by the shuttle of wee worlds, biotechnology.

Once on your way, you will find all you need to continue following your curiosity in the Appendix starting on page 167. Use it as your tool kit to dig as deeply as you wish. The Appendix includes: an explanation of *What Is Biotech?;* a unique description of *A Bacterial Bazaar;* a one-of-a-kind chronology on microbes in the environment; a glossary of key terms, defined in simple English; a list of expert sources, including addresses and phone numbers, with a brief description of each person's special expertise; and a listing of reference sources for your next steps.

A unique aid, this *BriefBook* will become even more useful as more genetically engineered microbes are released into environments throughout the world.

Before You Get Started

For most people, reading about science quickly turns into a definition derby. It's anything but stimulating and fun. And for most writers, writing about science presents a dilemma: either you define technical terms or you risk creating the cure for insomnia. Sure as sunrise, technical terms slow down readers and pester writers.

So reader and writer must strike a deal: the former must agree to use the glossary if the latter makes sure that it's written in plain English and doesn't require a Ph.D. to wade through. You will find such a glossary in the appendix. So if a term in this *BriefBook* — or any other book on biotechnology, microbes and the environment — doesn't make sense to you, check the glossary.

1 Infinitely Small, Historically Huge

"The role of the infinitely small is infinitely large."
Louis Pasteur

"We derive from a lineage of bacteria, and a very long line at that. Never mind our embarrassed indignation when we were first told, last century, that we came from a family of apes and had chimps as near cousins. That was relatively easy to accommodate, having at least the distant look of a set of relatives. But this new connection, already fixed by recent science beyond any hope of disowning the percentage, is something else again. At first encounter the news must come as a kind of humiliation. Humble origins indeed."
Dr. Lewis Thomas, "A Long Line of Cells,"
Wilson Quarterly, Spring 1988

As TV news statesman Ted Koppel said in 1984: "What is largely missing in American life is a sense of context. We measure the importance of events by how recently they happened." That's the problem with most of the coverage of microbial releases into the environment: The impression is that microbes are somehow recent inventions of scientists and thus new to the environment or, worse, that mankind's relationship with microorganisms just began.

The truth is that some genetically engineered microbes are indeed new creations that never existed before. But scientists did not invent microbes — indeed, they were around long before humans ever strutted out of the primal ooze. Perhaps, as Dr. Lewis Thomas muses, microbes "invented" us:

"For sure, I am fond of my microbial engines, and I assume they are pleased by the work they do for me.

"Or is it necessarily that way, or the other way round? It could be, I suppose, that all of me is a sort of ornamented carapace for colonies of bacteria that decided, long ago, to make a try at real evolutionary novelty."

Most writers who attempt to describe the history of microbes start with beer, wine, bread and cheese. On one level, that makes perfect sense: Human experience with microbes indeed began with a buzz, a baguette and some Brie. But that's *human* experience — history viewed in *human* terms.

Microbial history is another matter. And it matters a lot because it provides some badly needed context in which to view and understand the issues raised by releasing microbes into the environment. So where should we start our history of microbes? A lot closer to THE beginning than when humans had their first hangovers, that's for sure.

Life Before Oxygen

Earth was formed 4.5 billion years ago, give or take a couple hundred million years, and life or things alive arrived a billion years later. Those first alive things were microbes, specifically anaerobic — "non-oxygen-loving" — bacteria, and they had the place all to themselves for the next 2 billion years for a pretty compelling reason: Oxygen had not shown up yet.

Earth's early atmosphere, the product of many volcanic belches, consisted mostly of nitrogen gas, carbon dioxide and some water vapor — not unlike Los Angeles on an overcast day. Then, as Lynn Margulis and Dorion Sagan exclaim in their 1986 book, *Microcosmos: Four Billion Years of Evolution from Our Microbial Ancestors,* "the single most important metabolic innovation in the history of life on Earth" took place. The process of photosynthesis — during which water molecules get split and release O_2 or oxygen — evolved. This momentous event occurred in bacteria, the simplest kind of microbes. (You will learn the basic science of microbes — what they are, how they work and why their behavior outside is so poorly understood — in the next chapter, "Making Sense of Microbes.")

It's worth noting that some of this oxygen generated by bacteria floated into Earth's upper atmosphere and was converted into O_3 — ozone — thus forming the ozone layer that to this day prevents living organisms on Earth from being quick-fried to a crackly crunch by dangerous ultraviolet rays. In short, credit for the arrival on Earth of oxygen in two forms, the stuff we breathe (O_2) and the stuff that protects us from the sun (O_3), goes to aerobic — "oxygen-loving" — microbes, namely bacteria.

They Perpetuate Life

"The chemists of the living earth are bacteria. Their part in the cycle of life upon our planet is to effect a rapid decomposition of the constituents of dead animals and plants, and the products of their wastes, into simpler substances which are restored to the plant kingdom again to be rebuilt into living things. Inasmuch as some of these elements essential to life are limited in amount, this ceaseless activity of these industrious children of the living earth is essential for the very perpetuation of life upon our planet."

Arthur Isaac Kendall, *Civilization and the Microbe,* 1923

With oxygen around and the big ozone umbrella up, other more complex microbes made their debut roughly 1.5 billion years ago. More on them later, but for now it's clear that microbes have a history that makes ours look positively puny. It's a history that includes, among other things, making Earth habitable for all higher organisms.

We need microbes for our own survival; we cannot live without the deeds they do.

You see, microbes act as a kind of global janitorial service that literally cleans up or digests almost everything nature or mankind can make.

This fact has been recognized for decades, as the following excerpt from W.B. Benham's lecture, "Biology in Relation to Agriculture," delivered May 30, 1918, in Nelson, New Zealand, indicates:

"The general public may be pardoned if they do not recognize, or even know of, the vast utility of microbes. Life as a whole could not continue without their aid. They do not create life, perhaps, but they supply it with the necessary maintenance. It is by their means that the remains of dead animals and plants are cleared away and the organic substances stored up in them are returned to the air and soil in the form of inorganic matter, and rendered once more available for the food of plants."

This Bug's for You

As mentioned earlier, mankind's first brush with microbes — besides perhaps an inexplicable bellyache or flu — came when someone first quaffed a brew, sipped some wine or nibbled some cheese. You will find a brief synopsis of the entertaining culinary contributions of microbes in history starting on page 182 of the Appendix.

But this *BriefBook* is a story about the role of microbes in the environment and, specifically, humans' attempts to influence that role. The history of human microbial releases starts not billions of years back, but thousands. Now imagine yourself 2,300 years ago standing in a bumpy field of beans watching the handful of mysteriously rich soil slip through your weathered, stiff fingers back down to the ground from which you scooped it. The crop around you looks good, so your family will eat this winter. "That good," you think, as a broad smile reveals your teeth, badly ravaged by other mysterious microbial forces.

Magic Dirt

The Greeks knew it, the Romans knew it and other cultures doubtless did too: Soils seemed to have more fertility, more punch, after certain crops had grown in them. Theophrastus, a Greek alive 2,300 years ago, swore it was broad beans that left this magic behind in the soil. He hadn't a clue *why* — nor did anyone else — but he probably didn't care. It worked. Farmers who used soil from fields that last grew broad beans, or simply rotated another crop into such fields, produced more crop — maize, millet, wheat, whatever — and that's precisely what mattered most in early subsistence societies.

It took another 2,200 years before French chemist Pierre Berthelot, in 1885, suggested that some organisms in the soil might be able to convert or "fix" atmospheric nitrogen into a form that plants could use as fertilizer. The very next year German botanist Hermann Hellriegel said this "fixing" was the handiwork of bacteria that form nodules on the roots of leguminous plants.* In 1888, Dutch microbiologist Martinus Willem Beijerinck destroyed all doubt when he witnessed the bacterium *Rhizobium leguminosarum* nodulating peas. The news was out and soon confirmed in field trials conducted by John Bennett Lawes and his associate Joseph Gilbert at Rothamsted in England: Bacteria play the lead role in nitrogen fixation.

Europeans had taken the first big step in understanding soil microbiology.

Now, toting magical soils to so-so fields by the bucket or the wagonload proved that productivity could be passed from one field to others, but after 22 centuries of lugging dirt, humankind was ready for a more efficient scheme. Transporting the *bugs* — the microbial nitrogen-fixers themselves — made a lot more sense, for billions could easily be carried on "slants," an early culturing medium, inside small jars. Well, wherever a whiff of opportunity be, business noses will soon smell it. By the early 1890s, a German company, Höchst am Main, had begun selling the first commercially cultured *Rhizobia.*

In some fields, released *Rhizobia* did strikingly well; but in others, they failed miserably. And, once again, no one had a clue why.

"They're Everywhere..."

What those early soil scientists could not know is that other bacteria, many different species including *Rhizobia,* were already present in the soils they attempted to "inoculate." British microbiologist Dr. John Beringer, an expert in *Rhizobia* releases over the past 100 years, explains that the local microbes, though not as good at fixing nitrogen, are better at survival:

"The indigenous microbes essentially always win the battle."

*plants that bear their seeds in pods, such as peas, soybeans and broad beans

The "battle" here is for access to the roots of leguminous plants where *Rhizobia* must set up shop — or form nodules, to be precise — in order to begin fixing nitrogen.

Wait a minute. Where did these "indigenous" *Rhizobia* come from? How'd they get into farmers' fields before firms with the first *Rhizobial* brews? The answer to these questions illustrates a very important fact about microbes: They're *extremely* mobile. They are so small; they're so abundant — a teaspoon of soil contains *billions* of microbes; and they reproduce so rapidly, that they're *already* in most every environment on Earth.

World-renowned microbiologist Dr. Winston Brill describes the awesome mobility of microbes:

*"Some bacteria can easily survive and travel in comfort on a speck of dust wherever the winds blow them. Others move by rain, animals, insects, birds and plants — plants are literally covered with microbes, as are whatever fruits or seeds or grains they produce. If a microbe is stable and can withstand sunlight and temperature changes and so on and I dump some out in my back yard, that microbe will wind up in every single environment on this planet. It may not survive long in many, and no one will ever know it exists in many others, but I can say categorically it will at least visit **all** environments. They're everywhere."*

What's by far the easiest way for microbes to wind up in new places? Get sent there in first-class comfort. And that's just what the U.S. Congress did — unknowingly, of course — in what will no doubt remain the longest streak of deliberate releases of unknown microbes from all corners of the globe.

Franking Freeloaders

In fact, Congress shipped a veritable cornucopia of first-class microbe carriers — seeds — through the mails for *nearly 90 years.* Starting in 1839, the U.S. government eagerly pursued a policy of "franking" — sending something at taxpayers' expense through congressional mail — free seeds from around the world to any constituent who wanted some. The idea was simple: Try to make plants

from abroad crops in America. CSI's *BriefBook: Biotechnology and Genetic Diversity** picks up the story:

"In 1862, the year President Abraham Lincoln established the U.S. Department of Agriculture (USDA), the federal government distributed more than one million seed packages. The USDA did not start the distribution of seeds in the United States; instead, the USDA was started primarily because of seed distribution ...

"The seed franking frenzy proceeded apace, reaching its peak in 1897 when an astonishing 19 million packets of seeds were mailed. This free flow of free seeds, which the public obviously loved, continued until 1925 when Congress cut it off as a costly and inefficient way to disseminate, not to mention evaluate, seeds of potentially important new crops for the United States. No one knows how many millions of seeds were sent out, nor to whom, during the 86-year history of this popular program. Few useful records were kept."

The point: If microbes that interact with agricultural plants had not yet gotten to the States, Congress itself changed that — in a big

*See inside front cover for further information on CSI's second *BriefBook: Biotechnology and Genetic Diversity.*

way. For along with all those millions of seeds from exotic crops came unimaginable trillions of equally exotic microbes.

All this microbial movement took place before anyone had a clue about any of these foreign freeloaders — or knew that the greatest reason they weren't visibly changing the American landscape was the toughness of the indigenous microbes they encountered. Remember, "indigenous microbes essentially always win the battle."

First Planned Large-Scale Release

The microbe was a virulent bacterium and the insect it killed was the Japanese beetle, a pest accidentally imported into the United States sometime before 1916, when it was discovered in a nursery near Riverton, N.J. When the beetles ingest the bacterium, called *Bacillus popilliae,* it multiplies like mad in their blood, turning the beetle a milky white color — resulting in the name "milky disease" — and turning the beetle over to the great insect undertaker. In 1938, Bp, the bacterium, became the first microbial product registered by the U.S. government. The following year, a federal program was launched to speed up its use because the Japanese beetle found many climes east of the Mississippi River to its liking and was rapidly expanding its range.

Because no one had figured out how to grow the bacterium in culture, the deliberate release came in the form of ground-up grubs — immature beetles — mixed with chalk and sprayed as dust. According to Sam Dutkey, who headed the USDA's beetle eradication program from 1939 until 1952, some 230,000 pounds of bacterial dust were sprayed on 110,000 acres in 220 counties across 14 Eastern states.

That was one massive deliberate release — and it worked.

By 1945, milky disease was epidemic among the beetle populations of five states: Connecticut, New York, New Jersey, Delaware and Maryland. But, in spite of its success, this first federal and state effort in bioinsecticides was halted in 1953. Why?

Two reasons. First, the absence of an easy way to culture the bacterium made the natural treatment more expensive than chemical alternatives. Second, chemicals kill on contact or soon after, whereas bacterial insecticides take days or even weeks for full results. The plain truth is farmers — like millions of weekend gardeners — prefer quick results, namely to see offending bugs die.

This tortoise and hare difference in the delivery of results remains a problem to this day in the acceptance of bioalternatives to chemicals. Most people — especially farmers whose very existence often depends on dead bugs — don't like to wait for results.

Environmental Antibiotics

As it always does, war pushes scientific engines to their limits, and the pace of discovery quickens. World War II shifted scientists of all sorts — physicists, chemists, microbiologists and many more — into overdrive and what happened? Among many other things, the harnessing of nuclear power, the invention of synthetic insecticides and herbicides, and the explosive development of antibiotics all occurred during and as a result of WWII. One illustration suffices: Though purified penicillin — the forerunner of which came from a moldy melon in Peoria, Ill. — wasn't obtained until 1940 and the word antibiotic wasn't even coined until 1942, 20 U.S. factories were producing penicillin by 1943. By 1945, the number of antibiotics reached 30 — 150 by 1949 and 450 by 1953.

Seemingly overnight, antibiotics such as penicillin were miraculously saving thousands of lives. Chemicals were king, whether derived from nature or made by man. Agricultural pesticides, made by man, were seen as chemical miracles as well. They were, in effect, *environmental* antibiotics: chemicals that "cured" whatever ailed agricultural systems.

This image and the quotation below appeared in an advertisement in the June 30, 1947 edition of *Time Magazine.*

"The great expectations held for DDT have been realized. During 1946, exhaustive scientific tests have shown that, when properly used, DDT kills a host of destructive insect pests, and is a benefactor of all humanity."

In this light, the explanation for the swift switch from the first bacterial insecticide to the first synthetic chemicals becomes clearer.

What's more, only a few scientists at that time appreciated the interconnectedness in nature, and no one predicted that putting those chemicals into the environment would come back to haunt us. Indeed, as ecologist L.B. Slobodkin wrote in *Science* (11/88), ecology — the science of understanding those interconnections — was a field "so obscure in the early '50s that humorist Stephen Potter suggested 'oikology' (*sic*) as an ideally boring subject to be used in terminating romantic relationships."

So, starting in the late 1940s, the creation of chemical answers to nagging agricultural problems forced biological alternatives to fade into the background. But one legendary American scientist began wondering in the 1950s whether many of those chemical "answers" would not also create long-term environmental impacts.

She also predicted that bioalternatives to chemicals was the way of the future, but most only remember and revere her for what she opposed, not what she proposed.

Bugs vs. Bugs

Rachel Carson was ahead of her time in more than one way. She's a legend in environmental circles for single-handedly turning the world's attention to the harmful results of dousing various ecological niches with poisonous chemicals. First and foremost, Carson was a biologist. Perhaps her most fervent wish was that biology or "biotic controls" would someday replace chemical control of insects. That's not how she's best remembered. But it should be.

Carson clearly stated the direction she hoped her colleagues would pursue in the final chapter of her turning-point piece *Silent Spring* (1962):

"A truly extraordinary variety of alternatives to the chemical control of insects is available ... All have this in common: they are ***biological*** *solutions, based on an understanding of the living organisms they seek to control, and of the whole fabric of life to which these organisms belong. Specialists representing various areas of the vast field of biology are contributing — entomologists, pathologists, geneticists, physiologists, biochemists, ecologists — all*

pouring their knowledge and their creative inspirations into the formation of a new science of biotic controls." [emphasis added]

Carson included in "biotic controls" many non-chemical approaches — the use of insects against other insects, now called integrated pest management; the sterilization and release of males of an offending insect species; and even repellent bursts of ultrasonic sound, to name a few. But right at the top of her list was the use of *bugs* — microbes of various kinds and their products — against real bugs or insects. Again, she put it plainly:

"The new biotic control of insects is not wholly a matter of electronics and gamma radiation and other products of man's inventive mind. Some of its methods have ancient roots, based on the knowledge that, like ourselves, insects are subject to disease. Bacterial infections sweep through their populations like the plagues of old; under the onset of a virus their hordes sicken and die...Insects are beset not only by viruses and bacteria but also by fungi, protozoa, microscopic worms, and other beings from all that unseen world of minute life that, by and large, befriends mankind. For the microbes include not only disease organisms but those that destroy waste matter, make soils fertile, and enter into countless biological processes like fermentation and nitrification. Why should they not also aid us in the control of insects?"

Biotech Born of a Bug and a Toad

"Now at last, as it has become apparent that the heedless and unrestrained use of chemicals is a greater menace to ourselves than to the targets, the river which is the science of biotic control flows again, fed by new streams of thought."

Rachel Carson, *Silent Spring,* 1962

If the river of biology began flowing again in the 1960s, the streams that fed it could not compare with the flood that followed President Richard Nixon's full commitment of U.S. funds to winning the "war on cancer" in 1971. Billions of dollars began pouring into research on basic biology. Not since the top-secret Manhattan Project to produce the atom bomb in WWII and President John F. Kennedy's commitment in May 1961 to put a man on the moon before 1970 had so much money and talent been committed to an area of science.

And, as in the two earlier cases, something big was bound to happen. It did. And guess where?

In a bacterium. *Escherichia coli* to be exact.

In late 1973, Stanley Cohen of Stanford University and Herbert Boyer of the University of California at San Francisco chemically cut a gene out of a cell from *Xenopus* — the common toad — and spliced that gene into the genes in an *E. coli* bacterium. When they got the microbe to express the toad gene just exactly as if it were one of its own, Cohen and Boyer forever fixed their place in immortality; they had discovered "recombinant DNA." The press preferred to call it "gene splicing" and "genetic engineering."

And so, of a humble toad and a most basic bug, modern biotechnology was born. See *What Is Biotech?* on page 168. Biology — especially microbiology — had been changed forever.

With a sense of the huge history behind microbes behind us, let's get a grip on some basic microbiology.

2 Making Sense of Microbes

"Microbiology is so fascinating, but it's below vision. People know insects, but how can they imagine microbes? And here we're trying to teach the world about microbial releases! I think the thing that has impressed me more than anything is that getting a basic understanding of microbial genetics is so difficult."

Arthur Kelman, university distinguished scholar,
North Carolina State University

"Microbiology has undergone a revolution. In the past decade or so the techniques of molecular biology have revealed new, often unsuspected, relationships between genera and species of bacteria... Microbiologists will have to rewrite their textbooks."

John Postgate, emeritus professor of microbiology,
University of Sussex, in *New Scientist,* 1/21/89

O.K., confession time. How much do you know about microbiology? Unless you're a scientist, the honest answer is likely "very little." And that's no surprise. Microbiology is hardly a subject that captures lots of casual reading time; nor is it an easy subject to grasp.

To make things tougher, biotech — the common word referring to the techniques of molecular biology — is forcing even the experts to rethink what they know about microbiology.

Thus you arrive at the same frustrating position that many legislators, regulators, journalists, community leaders and concerned citizens find themselves in. With little or no science background, you're forced to evaluate the deliberate release of genetically engineered microbes into the environment. What do you do?

Dive in and do your best.

Some basic microbiology is what you need to make sense of debates on deliberate releases. You need enough knowledge to ask good questions of scientists and others who claim to know what they're talking about. This basic level of understanding is vital, because even the experts don't yet know the full story about microbes and know next to nothing about the behavior of microbes outside — microbial ecology, it's called.

What's a Microbe?

Microbes are cells that can live on their own as long as they have food and suitable climes.

Microbe is the common name for microorganisms, which are simply single cells, or clusters of cells, that can survive on their own.

Humans, animals and plants are basically big masses of highly specialized cells — kidney cells, eye cells, root cells, leaf cells, and so on. But, if you take any one of those cells and set it free, it won't get far, largely because of its specialization. A brain cell, for example, can only exist in the brain. Foot or hoof cells can't survive anywhere but in feet and hooves.

But microbes are an extremely diverse, massive group of free-living cells. Does this mean that microbes aren't specialized? No, it just means that microbial cells, more than anything else, specialize in independent survival. Microbes are Nature's masters of survival, her best adapters; indeed, so diverse are they that any niche in any environment around the world probably supports some kind of microbe — perhaps a few, even many kinds. Give a microbe the right niche and it will last a while — as long as it has food.

Bacteria: Life's Simplest Cells

All proks are bacteria and all bacteria are proks — it's that simple.

For roughly the first two billion years, the only living things on Earth were cells called prokaryotes (pronounced pro-carry-oat), a Greek word meaning "before nucleus." None of these cells had a nucleus, which is that center compartment in complex cells that you may remember from biology lessons long ago.

"Proks" are cells, but less complex cells. Turns out there's a handy synonym for prokaryotic cells: bacteria.

Eukaryotes: More Complex Cells

Then, 2.2 billion years ago, a major event in cell history — and thus a major event in microbial history — took place. More complex cells called eukaryotes (you-carry-oat) made their debut on Earth.

As you've probably guessed, "eucs* have nucs." This Greek name means "good or true nucleus" and eukaryotes do indeed have distinct nuclei, as well as other cellular machinery and complexity that proks just don't have. In fact, compared with proks, euks have tons more DNA: Proks have 1,000 to 5,000 genes, while eukaryotes lug around 200,000 to 3 million. Yet no one can say how much of the total DNA in any cell, prok or euk, actually forms distinct genes or how much is just extra DNA, sometimes called "junk genes." See box below *Does DNA = Genes?*

Does DNA = Genes?

DNA, deoxyribonucleic acid, is often called the stuff of life and it is always the stuff that genes are made of. But genes are specific pieces or lengths of DNA that, like chemical sentences, describe specific chemical products, mostly proteins. Now, just as you can put words together in a row that hardly makes a meaningful sentence, the "words" of DNA, called codons, can and often do come "in sense that order no make," if you get the point. As this is not a true sentence, codons don't always fall together in an order that makes genetic sense or, genetically speaking, makes a gene.

Genes are always made of DNA; but DNA does not always make a gene.

What every scientist can say, though, is that eukaryotic cells are what make up all organisms other than proks or bacteria. And that's all you have to remember for our story: Everything alive on Earth, except prokaryotes, is made up of one or more eukaryotic cells — plants, insects, fish, cheetahs and, yep, us.

The table on the next page, *The Euks Are Coming,* provides some examples of eukaryotic microbes and their fascinating skills.

*It's perfectly acceptable to spell both prokaryote and eucaryote with either a "k" or "c." For consistency we will use a "k" throughout this *BriefBook.*

THE EUKS ARE COMING

Some examples of eukaryotic microbes and the amazing abilities that make them so interesting to scientists and businesses follow.

Fungi: Fungi, including molds and yeasts, are euks that decompose things — just about anything. Cloth, leather, film, jet fuel, bread, wood — you name it, fungi can break it down. When they do, they produce by-products, making them both an enormous problem and enormous opportunity. Some produce toxins, such as aflatoxins that have carcinogenic effects at parts per billion; while others produce miracles, such as cyclosporine, the immune suppressant "wonder drug" that makes organ transplants possible.

In agriculture, one group of fungi — called mycorrhizae ("fungus-roots") — aids plants by living on their roots and helping them take in nutrients. But other fungi — like *Fusaria* molds — cause countless plant diseases, with more than 5,000 species known to attack plants of all kinds.

Some yeasts might play a role in reducing the greenhouse effect. Methane is a "greenhouse gas" that traps heat 20 to 30 times more effectively than carbon dioxide. Worldwide, 3.3 billion domesticated animals — and unknown numbers of wild animals — contribute more than 15% of all methane gas; the average cow belches or farts about 400 liters of methane gas *per day.* Certain yeasts can outcompete methane-producing bacteria in animals' guts and thus reduce the amount of methane that breaks out into the environment.

Thus, fungi do both "good" and "bad" things — and while 100,000 distinct species of fungi have been described, more than an estimated 200,000 await discovery.

Algae: Algae are single-celled eukaryotic plants that contain chlorophyll, the essential green pigment that allows the process of photosynthesis — deriving energy from light — to take place. While just 30,000 species have been identified, more than 150,000 species of algae are thought to exist. Algae represent one-third of the world's plant biomass.

Of the three divisions of algae — green, red and brown — the green is probably best known. It's the stuff you see in ponds.

Red algae produce agar, a principal ingredient in vitamin and drug capsules, cosmetics, and the slimy stuff cells are cultured on in petri dishes. Agar is derived from the cell walls of red algae. The Japanese have cultivated one red algae, *Porphyra* or "nori," eaten in soups and used as flavoring in many dishes, since the 17th century. The *Porphyra* harvested off Japanese and Korean coasts alone constitutes a $1.5 billion industry.

The brown algae are those seaweeds so conspicuous on shorelines around the world. *Macrocystis pyrifera,* the huge brown kelp that otters roll in off the California coast, provides "alginates" — compounds that make good thickeners and stabilizers for cosmetics, drugs, textiles, and food products like ice cream and cheeses.

Protozoa: Protozoa are single-celled animals — the smallest and by far the most abundant of all animals. Protozoa often feed on other microbes, mostly bacteria and some small algae. Thus, they might be useful in the control or maintenance of microbes that already have been altered and released.

But protozoa are best known as insect and animal parasites. Again this feature, though often the cause of debilitating diseases such as malaria, can be a potential advantage. A good example is *Nosema locustae,* "grasshopper sickness," the one protozoan registered as a pesticide by the EPA. It kills grasshoppers that consume $400 million worth of U.S. crops annually.

Tomorrow's Biotech Microbes

Because of their complexity, eukaryotes are tougher, much tougher, for scientists to study and manipulate — even with biotech. Nevertheless, only time stands between genetically engineered eukaryotic microbes and various environments around the world. For the most part, scientists are just starting to understand the role euk microbes play in the environment. Biotech tools help add to this understanding, so rest assured, genetically engineered eukaryotic microbes are coming.

So What Are Cells?

Cells are the basic units of all living matter. They are like microscopic chemical factories. Each has a cell wall enclosing its contents, namely the information or DNA that includes the genes needed to produce proteins and the machinery needed to manufacture them. In eukaryotes, the DNA is housed mostly in the nucleus, sort of like a main office, whereas in prokaryotes everything is simply inside the cell wall on one main factory floor.

Note that cells are *living matter.* This concept is very important. Cells are *alive.* Unlike commercial factories that may shut down at night, if cells shut down they die. Cellular factories are always in operation, exchanging materials with the surrounding environment and using them to produce whatever it is that the information in their genes dictates. Pulsating, moving, making and exchanging materials — these are the active words that describe cells.

The problem with any picture of cells, and therefore microbes, is that pictures are by definition static. Imagine a cheetah, the swiftest of all cats, pictured in a textbook; now see in your mind an image of a cheetah racing across an African veldt in hot pursuit of dinner for the kids. The first image shows you a cheetah; the second gives you a sense of "cheetahness."

So it is with microbes: Unless you see them in motion, you really can't get the right image — alive, pumping, writhing with "cellness" that, while not quite as moving to most people as cheetahness, is nonetheless a sight to behold. Especially when you realize that it is "cellness" that gives the cheetah "cheetahness" — and everything under the sun that's alive its "aliveness." After all, all living things are simply masses of specialized cells.

Microbes, like all cells, are alive, constantly interacting with their immediate environment.

Inside Information

Most of what is known about how microbes interact outside was learned in laboratories, not outside under the sun.

The next step in our basic lesson in microbiology is recognizing that most of what is known about how microbes interact outside was learned inside, in laboratories, not outside under the sun. To be sure, the field of plant pathology involves a great deal of study about how microbes in nature affect and interact with plants. But it does not

Life in the petri dish

include a focused study of how microbes interact with one another in countless environmental niches around the world, many of which do not even include plants. Dr. William Marshall, president of microbial genetics for Pioneer Hi-Bred International, explains why this is significant:

"If you think about where we learned the things that we know about microbiology, it's not been in nature. It's been in food microbiology, where a good

microbe is a dead one — kill 'em all off, salmonella and staphylococcus and so on. Or clinical pathology, where it's basically kill 'em off with antibiotics. Or we learned from industrial microbiology, which comes down to growing microbes in a tank, a fermenter. Until now, hardly anyone has ever had any interest in ecological microbiology, namely what goes on in nature. We don't even know the names of about 90% of microbes in the soil because who's ever cared about them — there wasn't a reason for it. Now there is."

What's more, until about 1980 when biotechnology began opening the black boxes that contain the secrets of microbial ecology, scientists studying what happens to microbes outside couldn't really conclude much. It was tough enough trying to tell one microbe from another. Marshall continues:

"Rhizobium meliloti is a species of bacteria commonly found on plant roots. Let's say, prior to 1980, that somebody says okay, I'm going to track the movement of my Rhizobium in the field. And you plant whatever, an acre or a thousand acres of alfalfa, and you put out all your Rhizobia. Now you isolate one, but you can't tell that Rhizobium from the next Rhizobium. You can look at it in the laboratory and maybe it has crinkles on it, little green crinkles, and if you've been looking at it for ten years you say yah, that's my guy. Number one, you can't be positive that's your guy. And number two, you can't be sure that it hasn't mutated into something else. But 99 times out of 100 you can't tell that Rhizobium from the next one unless you use modern genetic techniques that just didn't exist before 1980. So all those studies done anywhere in the world prior to 1980 I think have to be taken with a grain of salt."

Not Like Cats and Dogs

Scientists still have a heck of a time even telling one species of microbe from another. This point is very important in understanding microbes outside, as plant geneticist Dr. Peter Carlson points out:

"Most people have an image of species being pristine and sacrosanct. That's generally true with higher organisms — cows are cows, people are people and robins are robins. But with bacteria, species are really just groupings based on similar characteristics, not clearly distinct elements."

The reason species doesn't mean in bacteria what it does in higher organisms, as British microbe-watcher John Postgate wrote in *New Scientist* (1/21/89), is simple:

"Consider how biologists classify animals and plants. They observe and compare what the organisms look like, how they grow and what they do. Thus, a dog looks and behaves more like a wolf than a cat; all three are more like each other than a horse; all four are more like each other than they are like a frog, and so on. These relationships are borne out by detailed data on anatomy, development and physiology…

"When scientists came to apply these principles to microbes, things came unstuck."

Why unstuck? And why are microbes such tough little customers to tell apart, much less study?

It all has to do with that word "little" — it's hard to imagine just how little microbes are. But it's important to try because all too often when people discuss microbes in the environment — releasing, monitoring, tracking, evaluating them and so on — the microbes sound more like animals you can see, like wolves or rabbits. Nothing could be further from reality.

Wee Bits of Life…

To get a sense of just how small microbial small is, you've got to take a colossal leap into the minuscule and imagine yourself so small that even with outstretched arms you couldn't quite touch opposite sides of the period that ends this sentence. Now get this: 1,000 bacteria stretched end-to-end could fit inside your wide open arms inside the circumference of that period.

It cannot be overemphasized: The world of microbes is beyond human vision. Scientists can't see them without powerful magnification and that makes microbes very hard to study indoors and even harder outdoors. If scientists could observe bacteria feeding in a field of corn, say, as they can a bird, a grasshopper or a deer, we'd know a whole lot more about them. But they can't. So information is far more difficult to come by in the invisible microworld of bacteria, fungi and yeasts, than it is in the visible macroworld of lions, tigers and bears. Dr.

If scientists could observe bacteria feeding in a field of corn, say, as they can a bird, a grasshopper or a deer, we'd know a whole lot more about them.

Arthur Kelman drives home this point: "When you realize that 90%-95% of the microorganisms in a given soil sample don't even have names yet, doesn't that tell you something?"

This startling fact also reminds us that science is the study of what is *not yet known*. As a result, scientists often have to create terms that cover more of what they don't know than what they do. In the microbe story, one such term is "strain." Strains are groups within species that have certain characteristics in common — such as tolerance to heat or resistance to certain chemicals. Dr. Winston Brill explains strains:

"If I isolate a Rhizobium meliloti from a certain part of a field, it's one strain and, from another part of the same field, it's likely to be a different strain. They could be identical and thus the same strain, but there are countless strains because of all the different dynamics going on in any field. So, if you pick up a microbe from here and a microbe from there, chances are that they are not identical."

Ten thousand strains of *Rhizobium meliloti* alone are thought to exist — but nobody really knows how many thousands or millions of bacterial species, much less strains, exist on Earth.

…"Strains" To Study Outside

In late 1988, scientists at the University of Georgia's Savannah River Ecology Laboratory reported finding *10 times* more genetic diversity than had ever been recorded in a single species of soil bacterium. Dr. Michael Smith, one of the principal investigators, exclaimed to *Science News* (1/7/89): "Every single bacterium down there [in the soil] is unique, but each is unique in a way that matches very closely the differences in the environment."

90%-95% of the microorganisms in a given soil sample don't even have names yet.

Importantly, this does not mean that scientists cannot make some generalizations about microbial strains — people are all unique and yet doctors know certain things about people, like they survive best at 98.6 degrees Fahrenheit. It suggests, though, that the study of microbes outside supports Albert Schweitzer's insight: "As we acquire more knowledge, things do not become more comprehensible but more mysterious."

Truth told, no one really knows with precision what's going on among microbes in microspaces in the soil — or in microspaces anywhere outside a laboratory. Which is why, when you finally pin down the experts on how much we truly know about microbial ecology in nature, you get answers like these:

"Not very much at all,"
from Dr. John Beringer, world-renowned agricultural microbiologist;

"Hardly anything,"
from Dr. Winston Brill, world-renowned soil microbiologist;

"Next to nothing,"
from Dr. Arthur Kelman, world-renowned plant pathologist;

"Pitifully little,"
from Dr. Peter Carlson, world-renowned plant geneticist.

But Microbes Not the Focus...

Why should we let anyone release a microbe until more is known about microbial behavior?

When you realize that so little is actually known about microbes outdoors, you can't help but wonder: Why should we let anyone release a microbe until more, lots more, is known about microbial behavior in the trillions of microscopic landscapes across the globe?

For that matter, why would any investor want to risk big dollars on developing microbes intended for use outside, when so much work has yet to be done just to figure out what's going on out there?

The 1990s is likely to be remembered in biological history as the decade of debates about precisely these two questions:

- Are genetically engineered microbes safe to release into the environment on large scales?
- Do they work?

The '80s will be remembered as the decade of debates around the world over the *process* of biotech, not its *products*. Even the very event that started the stampede of investors to biotech in 1980 was process-based. On June 16, 1980, the U.S. Supreme Court decided to extend the reach of patent protection to living organisms. It did so because biotechnology had turned a living organism found in nature — heretofore an unpatentable "find" — into a man-made invention, thus the court considered it patentable. What captured the court's and the world's attention at that time was not the organism, but the powerful new technologies employed to alter it, loosely called gene splicing or genetic engineering. See *What Is Biotech?* on page 168.

...Biotech Is

That landmark 1980 case, *Diamond vs. Chakrabarty,* finds its place in history as the pivotal case for biotech, not for microbes. Few even realize that the organism in front of the court was a microbe — a bacterium that had been genetically engineered to eat crude oil. Even though this novel bug created the possibility of a cleaner, more effective way to clean up oil-soaked beaches, microbes simply don't grab people's attention the way visible things like whales and seals do. What did fascinate and/or frighten people was the *process* of biotech, or genetic engineering as most of the world thinks of biotech.

Proof for this key distinction — that it's "genetic engineering," not the resulting product, that makes the public pulse rise rapidly into its aerobic zone — is easily found. Microbes have been employed, albeit unknowingly, by mankind for many centuries and deliberately spread around soils and into environments across the world throughout this century. But never with fanfare — at least not until a deliberate release of a *genetically engineered* microbe was proposed in the early 1980s (see page 51).

It's "genetic engineering," not the resulting product, that makes the public pulse rise rapidly into its aerobic zone.

This is hardly a shock. After all, microbes do not yet make the stuff of riveting reading and arresting visuals that engender emotion as, say, three whales stuck in Alaskan ice do. Can you imagine, for instance, cozying up to the tube to watch, "Adventures in Microspace." The show opens with the electrifying sight of a teaspoon of soil. Even though in that teaspoon one could find 2,500,000,000 — that's 2.5 billion — bacteria, 400,000 fungi, 50,000 algae, and 30,000 protozoa, the intoned "billions and billions" just wouldn't have the emotional impact that Carl Sagan's voice does as his camera takes us off into the heavens.

Speaking of Alaska and network television visuals, another good illustration of the public's non-interest in microbes involves that terrible oil spill that blackened more than 1,800 miles of Alaska's coastline in March 1989. The following June, nine years after granting patent protection to that oil-eating engineered bacterium, the U.S. Environmental Protection Agency (EPA) — frustrated by the slowpoke clean-up of America's worst oil spill — deliberately released microbes with appetites for oil on some beaches in Prince William Sound. *Newsweek,* among the few to even pick up the story, wrote in its tiny story: "A new method for cleaning up massive oil spills will soon get a real life test as a result of the Exxon Valdez disaster." The method, well-documented since 1978, was hardly new; but the large-scale, open-air test was.

The microbes used to clean some beaches in Alaska were natives that had not been tinkered with.

Were the bugs the EPA used in Alaska genetically engineered? No. Do you recall hearing about their release?

You would have had they been. That would have made them big news, because genetic engineering is new and newsworthy. Microbes remain old and snoozeworthy — until they get genetically engineered, that is.

And that is the point that is illustrated again and again in this *BriefBook.*

But First a Few More Basics

Before we dive off the cutting edge of biotech into the deep end of release debates, we must master a few more basic points of science. Otherwise, as is often the case with debates around the world on deliberate release, the discussion may float off into mysterious heights where the air of understanding gets thin and eyelids grow heavy.

1. Biotech's Biggest Role: Revealer. The many players in biotech debates have different perspectives, but they have in common questions about how microbes work and live in the great outdoors. Everybody wants answers about microbial ecology — businesses developing microbial products destined for outdoor use; microbiologists driven by a desire to understand how the huge puzzle of Earth's microbial matrix is pieced together; regulators seeking systems that balance relative risks with relative benefits; concerned citizens demanding more direct say in what science does for and to our only home, Earth.

The secrets so many seek about microbial ecology have been locked in Nature's black boxes, and most can be unlocked only by, you guessed it, biotech. They weren't opened before 1980 because biotech wasn't widely available. Now it is, and herein lies the most important point that gets buried in biotech debates: By far biotech's most important and truly revolutionary role is not as the producer of products, but as the revealer of knowledge. Microbes live and work at molecular levels; biotech is literally on their level. Until the 1980s brought the application of modern molecular techniques to microbes in nature, scientists could not get down, so to speak, with microbes and study them on their own terms. Now they can.

By far biotech's most important and truly revolutionary role is not as the producer of products, but as the revealer of knowledge.

On this point, Dr. David Tepfer, an American-born French microbial expert, is emphatic: "You must remember that until only very recently, very few people were working on microbial ecology. It's only just now beginning. I think it's going to go very fast because there's suddenly a motivation for it. But it has been a totally neglected area — *totally.*"

2. Biotech's Real Power: Specificity. The most important reason biotech is rapidly changing microbial ecology into one of the hottest areas of modern research is its specificity. With biotech, scientists finally can know exactly what they've done after tinkering with the genes of any organism. Without biotech, no scientist can tell you with precision all the changes he or she has wrought.

Using the techniques of biotech, mutations or changes can be small and specific: one or a few genes added or subtracted at specific sites, all of which can be confirmed by various methods. Compared with the shotgun effects of general mutagenic techniques, which were standard practice before biotech and many of which are still widely used, biotech is revolutionary in its specificity.

Throughout most of this century, scientists have been using general mutagenic techniques — altering agents like chemicals, antibiotics, beta and gamma rays and more — to change or mutate the genes in microbes and other organisms, such as plants. They have released the products of these changes with virtually no regulatory oversight. With general mutagens, though, no one has a clue exactly what's been altered in the genetic code — not how big a change took place, nor exactly where, nor what any of the changes mean. If the product demonstrates the trait you hoped to mutate into or out of it — enhanced frost-fighting capabilities, say — then you figure your mutagenic technique worked. But you can only wonder how many other genetic changes took place too.

One good example of how biotech's power of specificity is literally building the foundation of microbial ecology is the advent of marker genes. As the name suggests, marker genes are genes put into microbes or any cells, that enable scientists to identify those microbes or cells among the billions of others outside once they have been

released. For example, scientists chemically cut the gene out of a firefly that makes it glow and splice that gene into the entire genetic recipe — the genome, as it's called — of a bacterium. They can then positively identify that bacterium when it glows like the firefly.

Marker genes are like genetic i.d. tags that allow microbes to be found in microspaces just as name tags allow pets to be found in macrospaces.

3. Bacteria Behave *Genetically* Very Differently. With all the talk about the movement of genes among organisms intended for use outdoors, one big point about basic microbiology is often missed.

Bacteria behave *very differently genetically* than all other organisms do.

For starters, bacteria do not have sex. "Well," you silently quip, "that surely makes them different from me." That's true, but it also makes them different from *all* other eukaryotic organisms that only exchange genes via sex, which is the mixing of genes from two sources. What's more, most higher organisms can only have sex with members of the same species. To be sure, some animals can mate with other species in the same genera, as is the case with tigers mating with lions; and some plants — like orchids — can even cross genera and still exchange genes. But, with all euks, gene exchange is a *sexual* process in which genes from two donors get mixed.

Amoeba porn flicks.

With proks, gene exchange is a one-way process without sex. Now you may think that bacteria are missing out by not having sex, but the important point is that you do not think that, as a result, genes are not moved around the bacterial world. The opposite is true. And it's important to understand why.

In the world of prokaryotes, which is without the limits of sexual barriers, genes move with far greater ease.

It turns out that bacteria, *because of* their simplicity not *in spite of* it, move genes around far more readily than do eukaryotes. Because bacteria are really genes plus a few other things just inside a cell wall, they are always just that wall away from new genes that may help them in one way or another, by allowing them, say, to quickly adapt to a new chemical that's been sprayed on them, or to crude oil that's been spilled on them.

Genetic exchange via sex is the only way genes move in eukaryotes. But, in the world of prokaryotes, which is without the limits of sexual barriers, genes move with far greater ease.

Bacteria Really "Get Around"

This brings us to the critical questions that must be asked when considering the release of genetically manipulated bacteria: Will the genes that scientists splice into bacteria be transferred to other, so-called "non-target" bacteria? And will novel genes never before a part of the prokaryotic world permanently alter it — perhaps in harmful ways?

Well, yes, maybe, possibly, potentially. And for two reasons.

1. They're Alive: When you release *living* organisms into the environment — bacteria, fungi, plants or animals — they can multiply and spread. Bacteria don't move far; indeed, they can hardly move at all. But they can move on other things — any other things. Inanimate things like chemicals and air pollution can of course spread, but they cannot multiply.

2. Their Genes Can Get Around: No gene added to higher organisms can be passed on to subsequent generations — unless it's added to sperm or egg cells. This means that genetic alterations of non-sex cells in higher organisms are by definition dead ends. The reason has to do with cell specialization; you recall that kidney, liver and brain cells or eye, hoof and tail cells can't be anything else. Likewise, eukaryotic germ or sex cells *only* exist to ferry genes to the next generation, via sexual mixing.

But bacteria can, through some really clever non-sexual tricks, get into each other's genes, so to speak, with much greater ease than the rest of the living world can. And, because of this fact, bacterial genes can get around even though bacteria themselves can't physically move much at all.

Bacterial Bazaars

"Our microbial ancestors made use of quicker ways for bypassing long stretches of evolutionary time, and I envy them. They have always had an abundance of viruses, darting from one cell to another, across species lines, doing no damage most of the time... but always picking up odds and ends of DNA from their hosts and then passing these around, as though at a great party."

Lewis Thomas, "A Long Line of Cells," *Wilson Quarterly,* Spring 1988

For the full story of Nature's prokaryotic Peyton Place, namely how genes move around the bacterial world, turn to *A Bacterial Bazaar* on page 171. For now, realize that because of the following five tricks and the clever cast of characters they involve, genes can move between and among bacteria much more easily than among all the organisms above them in the evolutionary tree.

Imagine for a minute that you can get so small that you too can cruise through a Bacterial Bazaar.

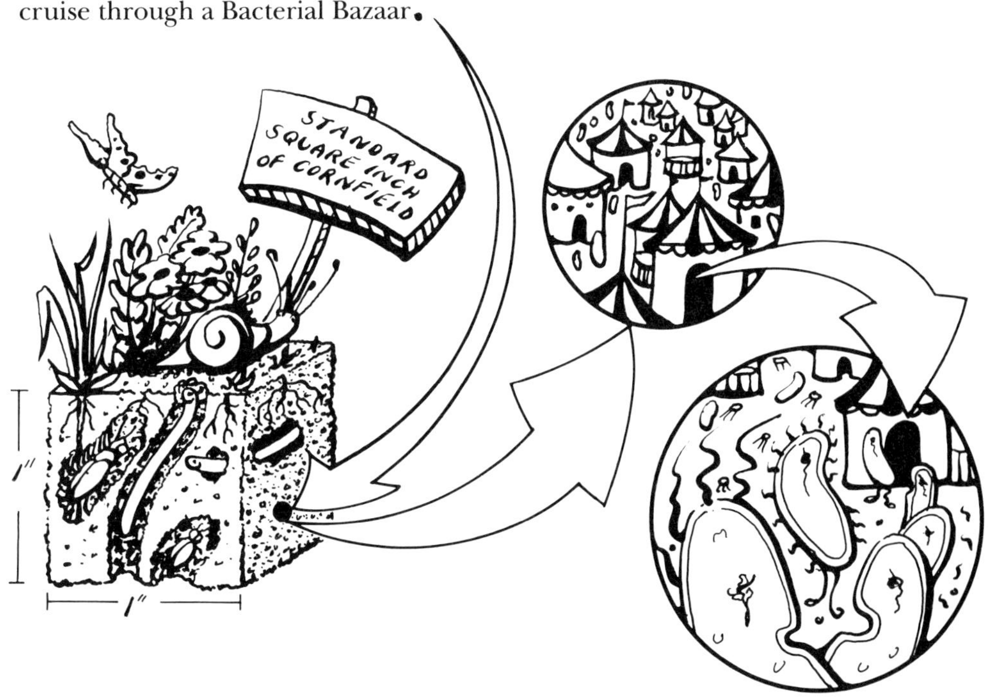

You are surrounded by writhing cells of all kinds and descriptions, with fluid shapes and watery, glistening colors, packed by the billions into tight quarters. Some are so close that they become connected for a time and genes move directly from one into another. Other cells suddenly "Pop!" — spewing genes all about that create a frenzy of activity around them. Other cells and suspicious-looking characters (viruses) snatch up these loose genes the way people grab money that accidentally drops from a passing armored car. And all this activity never ends, never even closes down for the night or to go on vacation. It simply goes on and on as it has for billions of years.

Now imagine that in trillions of microspaces around the world Bacterial Bazaars beyond your vision are operating right now.

The five known forms of natural gene transfer in Bacterial Bazaars are:

1. Transformation: Free or "naked" DNA enters directly into a bacterial cell in a straightforward process similar to the way it takes in nutrients.

2. Conjugation: DNA is transferred one-way from a donor cell to a recipient cell via direct contact in a downright steamy, physical act that involves a kind of gene taxi: plasmids.

3. Transfection: DNA gets a ride into a recipient bacterial cell from one of nature's cleverest creations — the virus.

4. Lysogenization: Similar to transfection in that it involves viruses, lysogenization is a more active moving of genes in response to stresses, such as being sprayed by toxic chemicals.

5. Transduction: An active process like lysogenization, but the DNA in this case gets carried along by viruses or plasmids.

Learned in the Lab

Like most of what is known about bacteria, these routes for gene transfers were gleaned under microscopes *inside* laboratories around the world. As a result, more and more experts are starting to aver, as microbiologist Sorin Sonea did in a fascinating article for *The Sciences* (8/88) called "The Global Organism: A New View of Bacteria:"

"Because bacteriologists have tended to study small numbers of specific strains completely removed from the dynamic bacterial stew that makes gene transfer possible, these processes have been undervalued and, in some cases, misunderstood. The fact is, bacteria could ill afford to contain so little genetic information if they did not have access to the genes of other strains."

To this day, leading modern microbiology and biology texts hold that "genetic recombination in prokaryotes is usually a rare event."* Accepted science says, even with all those funky tricks prokaryotes *can* use to move genes around Bacterial Bazaars, random mutations — simpler genetic changes caused by any number of potential mutagens, such as pesticides, herbicides, antibiotics, ultraviolet rays, you name it — mean more to bacteria's ability to adapt and evolve. Another of today's leading texts states: "Mutation is a far more important source of variability in bacteria than is genetic recombination."**

But there's a hitch: Random mutations *cannot* account for the speed with which certain traits whip through Bacterial Bazaars. Resistance to a certain antibiotic, for example, can arise in Australian bacteria one day, be found in San Franciscan bacteria the next week and in Parisian bacteria the week after. What explains this rapid transfer?

In part, that cast of creative characters that move genes through Bacterial Bazaars — plasmids, viruses, prophages, sex pili and so on.

Compelling new evidence for bacteria's high-exchange lifestyle came in late 1988 and again in early 1989. Scientists at two leading U.S. universities published strong evidence that pushed one of microbiology's oldest dogmas — that bacteria adapt to environmental stresses *after* they occur and then only because of random mutations — out into the hot light of re-examination. It now appears that, as the images of a Bacterial Bazaar suggest, bacteria may activate genetic changes or mutations that enable them to prepare for various environmental stresses *before* they occur. They may not just wait around until some random event hits them.

Easy to underestimate, these invisible little devils are. And there's more.

*See page 265 of *Biology of Microorganisms* (5th Edition), by Thomas D. Brock and Michael T. Madigan.
**See page 174 of *Biology of Plants* (4th Edition), by Peter H. Raven, Ray F. Evert and Susan E. Eichhorn.

Bacterial Bands in Global Gig?

Up front in this chapter we learned that microbes are single cells that *can* live on their own. But do they? Or, do they, as an ever-growing list of scientists now believes, actually work together to survive outside?

Welcome to perhaps the hottest debate in microbiology today: Are bacteria, as has long been the accepted rule, individual sorts fighting

like mad among millions of their own kind and untold billions of other microbes each only looking out for No.1; or are they actually social creatures toiling together in Bacterial Bazaars the world round? Sonea speaks for an increasing number of microbiologists when she claims:

"Local bacterial teams continually correct the mixture of their constituents by replacing less efficient strains with more appropriate ones. Although each strain possesses only the genes necessary to multiply and perform its specific enzymatic function, all the genetic information that exists in a local team is available, when needed... to any cell. Thus, each bacterium, itself a broadcasting station for hereditary information, is part of a much larger network."

Wow, imagine that: One giant gene pool, or ocean really, that bacteria bands — and potentially all individual bacteria — can dip into and take what they need to make them tougher in the global survival gig. That's heady stuff, and a far cry from single cells living the stereotypically "cell-centered" single's life.

Global Network of Bazaars

This possibility of cellular bands playing together also means more tough questions for those who want to release microbes with new genetic recipes into the environment. If whatever genes they add to the gene pool are available to any or even some of the other bacteria "in the band" — or "in the same Bazaar" or "in nearby Bazaars" and so on — that could mean trouble. It could also mean nothing — it all depends on the gene and the environmental niche it winds up in.

Furthermore, new evidence brought to the fore by biotech is bolstering the case for bacteria's bazaar ways. The May 1989 issue of

BioTechnology reported that recent research in France "demolished" another accepted notion in microbiology — that plasmids only move between various species of bacteria. A team of French and Israeli scientists used biotech to prove that plasmids can move freely even among very different *genera* of bacteria.

It's essential to emphasize that, because they are not themselves mobile and because of their short lifespans, bacteria only exchange genes with other bacteria in their immediate vicinity. Whomever they share any wee niche with is likely whom they will exchange genes with — if they exchange their genes or DNA at all.

Thus, perhaps the best image is a world of trillions of tiny Bacterial Bazaars, each occupied by bacteria from a given niche. How do all these countless Bazaars interconnect — for that matter, are they really connected at all? Great questions, but nobody has any good answers — yet. In fact, the whole subject of microbial gene exchange — especially outdoors — can truly be summarized as a field in which the more techniques like biotech reveal, the less we seem to know.

Big Stakes

Why is all this Bacterial Bazaar stuff so important? Because bacteria are the simplest organisms and thus the easiest for biotechnologists to manipulate. Biotech was, after all, born in a bacterium (see page 15). And ever since, scientists from industry, academia and governments around the world have been using all the tools from the biotech toolbox on bacteria. Then these scientists seek to test the results where they mean the most: outside. Thus, bacteria to date have dominated the list of organisms released into the environment.

Remember, bacteria are the stars of release dramas unfolding around the world.

Bacteria are the simplest organisms and thus the easiest for biotechnologists to manipulate.

To be sure, this is not to say that eukaryotic microbes — and of course plants and animals — are not themselves the subject of intensive biotech experimentation. They are. See *The Euks Are Coming* on pages 20-21. But euks are more complex and much tougher to manipulate. According to the same leading college text on microbiology* quoted earlier, the process of genetic recombination "has been too complicated to analyze in eucaryotes."

In the 1990s, the world will hear plenty about microbes, mostly bacteria and mostly because of biotech, moving outside. And because bacteria can shuffle genes about in unique ways with unknown frequency and effect, the stakes surrounding bacterial releases into the environment are high.

The 21st Century Technology

So which genes *are* biotechnologists introducing to that prokaryotic drama that keeps the whole world — from the rain forests, to the wheat fields, to the Alaskan waters and even our own back yards — balanced ecologically?

Well, now that you know a good bit about microbiology, you're ready to dive right into the deep end of deliberate release debates of the 1990s. But first let's look at a few of the 1980s debates to understand why the 1980s will be remembered as the decade of focus on the *process*.

That powerful process is biotechnology, the tools and techniques that 100 opinion leaders, polled by the Wirthlin Group in June 1988, bet will be *the* technology of the 21st century.

Now, time for some drama.

*See page 265 of *Biology of Microorganisms* (5th Edition), by Thomas D. Brock and Michael T. Madigan.

3 Going Outside: Releases in the '80s

"Olympics Canceled; Calgary Covered with Bacterial Ice!"

Imagine you are in charge of all skiing events at the 1988 Winter Olympics in Calgary, Canada. With the games just weeks away, your slopes don't have nearly enough snow to feature the world's super skiers. Italian sensation Alberto Tomba is on his way and all you have is a big headache.

What do you do? Pray for snow, yes. But what else?

Along comes a U.S. company selling a solution to your problem. You're all ears now. The secret to that solution is microbes — bacteria — that make it easier for water to freeze at higher temperatures. The company officials explain that the bugs in the brew were grown in large fermentation tanks — just as beer and soy sauce are — then freeze-dried into pellets and finally zapped with radiation.

"Why'd you nuke 'em?" you ask nervously. To make sure the microbes cannot multiply once released, explain the company's experts. Besides, they assure you, these bacteria are no different from ones Nature has already put on your slopes — although obviously in very small amounts. A little joke you don't find amusing. The games are hurtling toward you now like so many downhill racers and you desperately need new snow.

Like most people, you don't understand bacteria, much less the effects radiation has on them. But, you figure, "Since they're already in Nature, why not?"

So you approve spraying the bacterial snow a few days before the games. Suddenly, your phones start ringing off the hook. It's the press — the entire world's press — acting on information from an anonymous activist, demanding a statement about the nuclear mutants being loosed by the billions that could cause all of Calgary to freeze — possibly some Canadians as well — and ruin the games.

You have a very big headache now.

Seems far-fetched, this scenario. But only the headline and panicked media reaction are.

Here's what actually happened. Snomax Technologies, part of Eastman Kodak's Bioproducts Division, saved Olympic organizers and Calgary officials some serious headaches by selling them Snomax® — a snow-making solution based on the bacterium *Pseudomonas syringae.* In fact, Snomax now helps 200 ski slope operators around the world — Japan, Australia, Europe, the United States — fill in those sections on the slopes that Mother Nature misses.

"Scientists Fight Frost with Natural Microbe!"

Now imagine the reverse scenario. You're a scientist and you've managed to delete a gene from the same bacterium that Snomax is based on and thereby decrease its ability to form frost. But your goal is to *prevent* the formation of frost and save farmers around the world the huge sums they lose each year because of frost damage — estimated at $11 billion to $14 billion.

Here again, Nature is way ahead of you. *Pseudomonas syringae* without the gene that enables ice to form — the so-called "ice minus" bacterium — can be found virtually any place the "ice plus" version can be found, namely, wherever frost forms on plants. After all, Nature was adding genes and deleting genes — recombining genes — in countless organisms long before she cooked up the idea of humans.

Thrilled by your "ice minus" bacterium, you name it Frostban® and make plans to spray some frost-sensitive plants, strawberries in this scenario, outside to see if it works. The community, farmers and the media are all ecstatic about your innovation, especially since it might replace the need for polluting smudge pots that burn throughout California nights when early frost threatens.

Fame and fortune will surely flow from your ingenious idea.

This scenario is also true except for the part about the warm reception for Frostban from the media, the community and farmers. Fortune didn't follow, but fame certainly did.

In what became the most widely publicized experiment in the history of biology, Advanced Genetic Sciences (AGS), a small biotech firm based in Oakland, Calif., was permitted by federal regulatory agencies to take Frostban outside and spray it on some strawberries in a field in nearby Brentwood on April 24, 1987. This seemingly simple procedure was covered by all the U.S. television networks and many stations from foreign countries. It even featured one of the world's first acts of "ecotage" — the destruction of property with the goal of saving the environment.

AGS had first asked the National Institutes of Health for permission to release Frostban back in 1982; five years passed before the company could test outdoors bacteria that Nature herself already manufactures and distributes widely. Ironically, AGS was also the creator of Snomax — the snow-maker and Olympics-saver; it had sold the product to Kodak to focus on agricultural applications of Frostban.

Why the night-and-day difference between the reactions to these two microbial products? Two very potent words: *Genetically engineered.* Frostban is and Snomax is not.

Minus and Plus = 1 Big Headache

Both forms of *Pseudomonas syringae* bacteria, ice minus (Frostban) and ice plus (Snomax), already occur in abundance in Nature. Both had been altered by scientists previously — either by isolating them from natural strains and growing them up in cell cultures or by mutagenic techniques* that, compared with biotech, are as precise as fishing with a shotgun. And both had previously been released into the environment.

But, clearly, had someone been so inclined, the Olympics public relations staff could have had major headaches with what obviously can be sensationalized. For instance, no one can say for sure that the bombardment of radiation didn't cause unknown mutations in Snomax that were turned loose into the Bacterial Bazaar with unknown consequences. Even "dead" bacteria contain genes that can be moved around the Bazaar in those creative ways we learned about earlier. (See *A Bacterial Bazaar* on page 171.)

But Snomax became no more than a small special-interest story at the '88 Games, dwarfed by more compelling stories like the sparks between Katarina Witt and Alberto Tomba.

Yet tons and tons of ink were put to page around the world, and the entire U.S. regulatory apparatus in Washington, D.C., was caught without a clue when scientists asked to test ice minus bacteria outside. Amazingly, it's a story that won't go away, as typified by the following quote from "Opening Pandora's Box," an article in the British publication, *The Ecologist,* published 29 months after the first release of Frostban:

"A U.S. firm has recently released a genetically engineered version of the bacterium Pseudomonas syringae... and it is intended that this mutant strain ('Frostban') should displace the 'damaging' natural one from the environment. This is an alarming prospect..."

*A changed or altered gene is a mutated gene, so mutagenic techniques are simply ways to alter genes. Chemicals, radiation and the set of techniques in the biotech toolbox shown on page 168 are all mutagenic techniques.

Why does the Frostban story linger? Because biotechnology makes so many people uneasy. It touches nerves. A creation of "genetic engineering" was going to be deliberately released into the environment. The United States, and the world really, simply was not ready for that.

Why is it that gene splicing specifically and biotech in general stir people so?

Yesterday Asilomar, Today the World

Soon after gene-splicing was born in a California laboratory in 1973, a group of forward-looking scientists began to wonder aloud whether they could fully appreciate its possible impacts — both inside the laboratory and outside — should something inadvertently escape. As a result, on February 25, 1975, these scientists convened the now-famous conference in Asilomar, Calif., to consider placing controls on the cutting and pasting of genes to and from various unrelated organisms — at least until more was known. The meeting started scientists and, in turn, the U.S. government down the tracks of regulating biotechnology (see page 111).

Those 100 scientists also sent a clear message from Asilomar to the world: Biotech was big, perhaps too big for even such big minds to fully grasp, much less control.

Thirteen years later, Dr. David Kingsbury of George Washington University Medical School, who has been a central player in biotech debates from the beginning, looked back at Asilomar:

"At the time, we weren't even out of Vietnam and a lot of the scientists were living on campuses defined by turmoil. Faculties had changed philosophies to respond to the mindset of the times — being very socially responsible was ***the*** *thing to do. All of a sudden, gene splicing became reality, suggesting far-reaching implications, and the scientists involved wanted to lead society down the right path.*

"The scientific community, which had no notion whatsoever about regulation, bought into Asilomar's cautious message as a concept without ever thinking it out, really. Those few forward-looking scientists — like Stanley Cohen and Bernard Davis — who said, 'Folks, we're sending the wrong

message here; recombinant DNA is part of nature; it's not so new and it's not itself dangerous,' were seen as old curmudgeons who weren't socially aware."

Thus, up went a red flag over biotech for all the world to see.
For the next five years, the wave of concern Asilomar initiated swept across the United States and around the globe, with its main focus on the complete containment of both the products and the processes of biotechnology. Could biotechnology be bottled up and kept safely indoors, or might monster mutants and deadly germs escape, run amok, wreak havoc and take over?

Good News is No News

As it turned out, all the early experience with biotechnologies, in countless experiments in thousands of laboratories around the world, indicated that the initial concerns about "it" — the process of biotechnology — and "them" — the products of biotechnology — were vastly overstated. Indeed, to this day, in the now millions of experiments conducted worldwide that involve one or more techniques from the biotech toolbox, no harm has come from biotechnology. But this telling track record rarely gets told.

Safety, of course, is not news.

Not only have biotech processes not caused harm to humans; no biotech product, and specifically no genetically engineered microbe, has caused harm to humans or to the environment. Dr. Roy Curtis, one of the Asilomar architects of early containment rules for biotech in the United States, emphasized at the first international conference on the release of genetically engineered microorganisms — dubbed REGEM 1 — held in Cardiff, Wales, in April 1988:

"I have yet to hear a single specific example of a possible harmful consequence in the use of any genetically modified microorganism designed to achieve a beneficial purpose."

Millions Escape

Does this mean that mutated microbes, regardless of the techniques used to alter them, haven't escaped? Absolutely not.

Doubtless they have, from all but the most carefully controlled labs, P4 facilities as they are known. Curtis explains:

"At the height of the early rDNA controversy, in the mid-1970s, we did many studies to try to assess the number of microorganisms that can inadvertently escape a physical containment facility, depending on the level of containment used. Forget the exact numbers, but roughly speaking, from a P1 lab, doing the normal sorts of microbial genetics and molecular biology, you probably release on the order of 10 to the 8th or 9th [power] microorganisms per day per researcher. P2 facilities reduce that a thousand- to ten thousand-fold, but it's only when you get to a P4 laboratory that you can effectively preclude release of microbes.

"How do they get out? Mainly on people — on your hair, your clothes and so on."

So millions of microbes actually do escape from all but the most tightly controlled settings. And, since all the microbes that will be deliberately released into the environment — the focus of our story — do not require P4 or the tightest containment, they escape all the time. In fact, microbes destined for outdoor release get tested under the *least* stringent protocols — in greenhouses or other P1 facilities, which are to microbial containment what a colander is to water containment: more like sieves than safes.

Put plainly, any microbe that has made it to the greenhouse for testing has *already* gotten outside. But the public perceives that greenhouses are still indoors and perception, not reality, perpetuates notions that we "know" aren't real — like airplanes are a dangerous way to travel and products of genetic engineering are dangerous *once they get outside.*

Giving Biotech the Business

By the early 1980s, biotech businesses were busily selling biotech's promises to eager ears on Wall Street and in newsrooms. The climax came on October 14, 1980, when Genentech set a record that still stands on Wall Street: Just 30 minutes after its stock was initially offered to the public at $35 per share it shot like a meteor to $89. No other initial stock offering in history had peaked with such velocity. Speaking for many opinion leaders, John Naisbitt, in his 1982 mega-seller *Megatrends,* couldn't contain himself:

"Gene splicing: more important than atom splitting — unless, of course, we blow ourselves up... We will be able to to create a second 'green revolution' in agriculture, to produce new high-yield, disease resistant, self-fertilizing crops."

But as Nancy Pfund, partner at Hambrecht & Quist Venture Partners, said in early 1989:

"The revolution in products that biotech promised never arrived, except for a few cases in medicine. Even so, until two years ago, biotech remained a darling of the investment community because it's the perfect embodiment of the American Dream."

So the public was promised a revolution that didn't arrive and that, according to Dr. Peter Carlson, a pioneer in the development of genetically engineered microbes:

"Makes people mad. People don't like revolutions because they connect them to bad things. Look at what Mao's revolution did to China or Lenin's to Russia. Revolutions change people's whole world and few like that. So they get excited and let down and that doesn't help the perception of biotech at all. We in businesses are largely responsible, caught in our own rhetoric trying to sell stock."

As one biotech stock-watcher told *Changing Times* in June 1987, "Biotech company officials are spouting projections that have no reality. They whip up the public's imagination every time they rinse out a petri dish." By continuing to hype biotech's power and promise and then turning around and saying, right before releasing genetically engineered microbes, that "they're no different from what's been released before," businesses have confused the public.

U.S. House of Representatives Agricultural Committee consultant, Skip Stiles, describes the result:

"The public hears big chemical and energy companies saying, 'Hey, c'mon, trust us, biotech's safe and clean,' and the public just mutters, 'Yeah, right.'"

The Supreme Tailor...

Still another sensitivity button was stitched into many minds on June 16, 1980, when the U.S. Supreme Court, in a tight 5-to-4 decision, held that a bacterium whose idea of the perfect meal is crude oil — created for General Electric by microbiologist Ananda Chakrabarty — was *patentable.* Plain as vanilla it was: The living products as well as the "revolutionary" processes of biotech could be owned and legally protected just as VCRs, cellular phones and personal computers are.

Most people still do not understand the role patents play in making information *more* freely available to the public, not less so, but that's not what mattered anyway. Most simply felt at a gut level that there is something odd about patenting living organisms.

The key point for the release story is this: Biotech suddenly became much bigger business — and covered by the media that much more — because the highest court in the United States had given investors around the world a green light to patent biotech products and processes.

...Sews Up the Deal

By the early 1980s, the deal had been sealed: Biotech had been sold to an unaware public as "Super Science." Faster than speedy scientists could keep up with, able to leap long-standing Wall Street records in a single day; made more powerful by potent patent protection — "Look, it's Super Science! It's BIOTECHNOLOGY!!!"

Put it all together: Concerned scientists, eager entrepreneurs, vigilant reporters and just enough justices perfectly prepared the public. When someone dared to request permission to *deliberately release a genetically engineered mutant into the environment,* an avalanche of anxiety would fall.

Someone did just that.

An Angst Ambush

Picture in your mind for a moment the final scene from the movie *Butch Cassidy and the Sundance Kid.*

As they have so many times before, Butch and Sundance bolt into the sun with guns ablaze, assuming that they'll reach their horses,

ride off into the sunset and carry on with other risky experiments. We know they don't have a chance, because they grossly underestimated what was out there — hundreds of Mexican federales waiting to end the movie by ending Butch and Sundance. Guns roar; freeze frame.

Something similar happened in August 1982 when Steven Lindow, a big, likeable, bear of a scientist from U.C. Berkeley, became the first to ask permission to *deliberately release* genetically engineered microbes into the environment. He too didn't have a chance, and he too didn't have a clue what was waiting out there ready to ambush him — the avalanche of angst poised to bury the first person to push all those sensitive biotech buttons in the public's mind.

And, if Lindow was Butch, then AGS was Sundance, because they were right behind Lindow with their request to the NIH to test Frostban, developed from Lindow's technology, which they had licensed from the University of California.

Freeze frame on frost fighters — or any other genetically engineered microbes — going outside for the next five years.

The *Five-Year Freeze-Out* on the next page is a chronological summary of the ice-minus saga, the most publicized microbial release in history. To tell the full story would require a not-so-brief book of its own. Suffice it to say that Lindow and AGS paid the price that comes with massive media scrutiny while the U.S. government tried to get its regulatory act together for biotech in general and deliberate releases in particular. As you will see in Chapter 6, "Regulating the Revolution," starting on page 109, that federal act is still suffering from mixed reviews.

An angst avalanche came down on the first scientists to ask permission to release GEOs.

FIVE-YEAR FREEZE-OUT

1982 Dr. Steven Lindow, a plant pathologist at the University of California at Berkeley, uses biotech "scissors," chemicals called restriction enzymes, to snip the gene responsible for ice formation out of *Psuedomonas syringae.* These "ice minus" bacteria, it is hoped, will outcompete "ice plus" bacteria for space on plant leaves and thus prevent frost formation and damage.

August: Lindow asks NIH to let him test his genetically engineered ice minus bacteria outside.

Dr. Trevor Suslow of Advanced Genetic Sciences (AGS), a friend of Dr. Lindow, asks to test Frostban outside to determine its commercial potential for AGS.

1983 NIH unanimously approves Lindow's release request.

September: Jeremy Rifkin of the Foundation on Economic Trends sues to stop Lindow's release, claiming the United States doesn't have adequate regulations to review deliberate releases of any organisms. Rifkin was right; see page 114.

1984 U.S. District Court Judge John Sirica — the famed "maximum John" from Watergate hearings — blocks release because NIH failed to properly review potential environmental impacts of Lindow's federally funded research.

June: NIH's Recombinant DNA Advisory Committee (RAC) unanimously approves AGS's proposal to release, putting NIH Director Dr. James Wyngaarden in an awkward position: If he allows AGS to proceed, he will set a dual standard for private research and NIH-funded research.

Rifkin immediately files suit to stop AGS.

October: EPA announces to everyone's surprise that it will require notification for all small-scale releases of genetically engineered microbial pesticides. (See page 116.)

1985 Lindow and AGS comply, and send EPA notification of their intent.

March: U.S. Court of Appeals refuses to lift injunction against Frostban test, saying that NIH has not fully considered possible environmental impacts. At the same time, one judge criticizes Rifkin's use of the courts to delay, saying: "The use of delaying tactics by those who fear and oppose scientific progress is nothing new."

November: EPA approves Frostban release and Rifkin files suit, again averring inadequate environmental review.

December: California Department of Food and Agriculture approves Frostban test.

1986 Monterey growers begin to react, fearing public response and reduced sales. Monterey residents are upset that they have had no say in proposed release.

February: Monterey County Board of Supervisors, responding to national and local pressure, bans tests for 45 days.

The press learns that AGS conducted unauthorized outdoor tests of Frostban in February 1985 on the roof of its building in Oakland.

March: Because of AGS's unauthorized tests, EPA fines AGS $20,000, the maximum; Congress holds hearings; and Monterey County supervisors vote one-year moratorium on tests.

April: Local environmental groups protest in front of AGS building.

May: Having learned from AGS's mistakes, Lindow and U.C. officials go to Tulelake, Calif., to discuss proposed release there with residents.

EPA approves Tulelake test. Rifkin tries to block test by demanding that EPA require liability insurance for those releasing altered microbes.

June: Upon further review, EPA clears AGS of most charges and reduces fine to $13,000.

August: California Superior Court Judge A. Richard Backus issues temporary restraining order against U.C. release.

1987 Local farm associations in Tulelake oppose U.C. release as rumors circulate that Idaho potato farmers have prepared labels reading "Idaho Potato Grown in Mutant Bacteria-Free Environment."

AGS chooses Brentwood in Contra Costa County as proposed new test site for Frostban and, as a result of extensive community relations work, local farmers and officials come out in favor of tests.

February: EPA gives final approval for AGS test; California Department of Food and Agriculture and Contra Costa County Board of Supervisors approve as well.

April: Sacramento District Court Judge Darrel Lewis refuses to stop AGS test.

AGS begins Brentwood test. Scores of journalists from around the world watch as Dr. Julie Lindemann, dressed in a protective "moonsuit," sprays Frostban on an acre of strawberry plants. Vandals rip up strawberry plants, which AGS replants.

Lindow's Tulelake potato test begins.

May: Tulelake test site is vandalized by Earth First! activists and U.C. replants potatoes.

December: Second AGS test delayed when vandals spread rock salt and ammonia on site. Vandals call themselves "Mindless Thugs Against Genetic Engineering." Test finally goes forward, this time without moonsuits, which had never been required in all the years of testing ice minus until it was genetically engineered.

1988 November: Tests show that treated potato and strawberry plants suffer significantly less frost damage than untreated plants.

1989 AGS drops commercial development of genetically engineered Frostban to focus instead on other projects; Lindow continues his research.

Global Truths

The ice-minus story reveals some truisms about releasing genetically engineered products into the environment — truisms that apply most anywhere in the world. Ask any of the scientists and businesses, such as Lindow and AGS, who have run the gauntlet of media, regulatory and public scrutiny, how painful it is to learn these truths the hard way.

River of Angst Runs Deep: Never underestimate the latent anxiety about biotech — it's a river that runs deep, flows steadily and can rise rapidly at any time.

Without some basic background in both history and science, the public is, in effect, dropped on top of the biotech mountain and told not to worry about the dizzying heights. Scientists, on the other hand, have climbed the biotech mountain step by step and get quickly frustrated when others can't see that the route up is basically safe.

All the public can see is that, should anyone slip, it's a long way down.

Mixed Messages Matter: It will be some time before biotech will come down to reasonable levels from the supernatural heights it claims in many minds. No wonder. First the public was told by John Naisbitt in his 1982 *Megatrends:* "Gene splicing is the most awesome and powerful skill acquired by man since the splitting of the atom." Then, in 1989, Monsanto Senior Vice President Howard Schneiderman told *California Farmer* magazine: "A thousand years from now, when many of today's other technologies—microprocessors, robots, lasers—are old stuff, biotech will still be at the very center of much that we do."

But, come release time, the public is told that biotech is "part of a continuum" and "no different from what Nature already does." The result, according to the National Audubon Society's Maureen Hinkle, who has tracked biotech since its beginnings:

"There's an unfocused rage out there that biotech stimulates; people just don't want to be led down the garden path. It's impossible to predict where it will rise. But it's there."

The point: On the slippery slope of imagination, only knowledge provides footholds.

No Clear Answers... Yet: "Predictive ecology" borders on being an oxymoron because definitive answers about microbial behavior outside don't exist yet. This means that when the public asks simple questions about microbes, scientists' sentences start with "Tests in greenhouses indicate..." or "Nothing in the literature leads us to believe..." Not comforting stuff.

What's more, community leaders demand guarantees of safety, environmentalists require reliable predictions about what a given microbe might do outside, and scientists will not *be able* to deliver either. Indeed, scientists will only be able to give estimates — and ironically these can only be based on field data gained through, yep, deliberate releases. All of this points to the real dilemma for regulators considering releases: The best information with which to evaluate them will only come *after they're released.*

Catch 22, it's called.

The Ol' End Run

What's the surest short-term way to deal with the river of anxiety over gene splicing? Avoid it.

That's precisely what some biotech firms have done: sell like mad the fact that they won't be ambushed as AGS was because their products don't result from gene splicing — at least not yet. Let the AGSes and the CGIs — see next chapter — have the hot light of fame for being the first to release genetically engineered microbes. Simply sell your in-house biotech prowess and keep it "in-house" until the public and regulators get used to letting biotech products outside.

A good example is Ecogen Inc., a Pennsylvania-based biotech firm founded in 1983 to develop biopesticides. Ecogen's strategy — like that of just about every other company pursuing pesticidal microbes — is based on Bt, *Bacillus thuringiensis,* the well-known bacterium that kills bothersome bugs (see page 65). The difference is that Ecogen is isolating naturally occurring strains of Bt — from an in-house library of Bt that includes 4,000 strains —and then souping them up in some way by using well-known mutagenic techniques* that

*See the bottom shelf of the biotech toolbox on page 168.

have one very important thing in common: None is seen by regulators as biotechnology.

Does this really make a difference in safety when it comes to field trials? That all depends on the characteristics of the organism that's released and the specific environment that it's released into — not on the techniques used to alter it.

Does it matter as far as regulators, the press and the public are concerned? You bet.

Because Ecogen's first generation of products is "natural," they hardly get regulated at all. Since June 1986, when Ecogen was granted permission to field test a genetically altered *live* microbial pesticide without a permit — the first non-biotech, live, altered microbe to be released — the company has conducted more than 400 separate field trials of live microbes designed to kill various pests. All with virtually no public complaint, few regulatory requirements and, as a result of these first two facts, far less expense than anyone trying to release a "genetically engineered" microbe.

Are Ecogen's products, and especially the genetic mutations made in them, far better characterized and understood than AGS's ice minus? Not at all. In fact, in view of biotech's elegant specificity, *less* is known about them. They're just not genetically engineered and thus little attention is paid to them.

Dead Biotech Bugs Kill

In one of the cleverest regulatory end runs ever, Mycogen Corp., a San Diego-based biotech enterprise, devised a way to house the hype of genetic engineering in another package that doesn't arouse regulators: dead bacteria. Here's the trick: Scientists first splice whatever genes they want into live bacteria. Then, with a unique combination of heat and chemicals, they kill those bacteria, thereby encapsulating the toxin produced by those added genes in dead bacteria. These killer killed bacteria are sprayed onto infested plants and there it is: a dead bacterial delivery system. Mycogen calls its patented handiwork CellCap®.

In 1985, Mycogen scientists CellCapped, if you will, a well-known caterpillar-killing toxin from Bt and received permission to test the resulting bioinsecticide, called MVP, the very same year — the first genetically engineered biopesticide to receive the Environmental

Protection Agency stamp of approval. At the same time they isolated a new strain of Bt that's especially effective against the Colorado potato beetle and called their variety "San Diego." The CellCapped version of this beetle-killer, called M-One Plus, was approved for testing in 1987.

Once again, the EPA had little problem with MVP or M-One Plus because they not only rely on the already familiar workhorse Bt, but they also find their way into the field in *dead* microbes — not *live* microbes.

Do not be misled: Mycogen and Ecogen and everyone else developing bio-alternatives to chemical pesticides have their sights set on eventually selling *live* microbes with many different kinds of genes spliced into them. Other strategies, however clever and appealing, are short-term. As Mycogen Chairman and CEO Dr. Jerry Caulder likes to say of his CellCap end run:

"We found a technological solution to a social problem. That enabled us to get outside a couple of years ahead of our competitors."

Mycogen, Ecogen and others saw the writing on the wall: Getting *live* genetically engineered microbes outside into the environment is an expensive and unpredictable process.

Stuck on the How

Try as many people did throughout the 1980s to shift the focus away from the process of biotech to its potential *products,* nothing worked.

In early 1987, the National Academy of Sciences drafted one of its most distinguished members, Dr. Arthur Kelman, a plant pathologist then from the University of Wisconsin at Madison, to chair a committee that would write a brief "white paper" on biotech and deliberate releases. One of the first things "Introduction of Recombinant DNA-Engineered Organisms into the Environment: Key Issues" made clear was that the process of biotech does not add new risks to the products created by it. The Kelman report said it this way:

"Assessment of the risks of introducing R-DNA-engineered organisms into the environment should be based on the nature of the organism and the environment into which it is introduced, not on the method by which it was produced."

The Kelman report was widely criticized for various reasons — its hasty production, its unorthodox, simple style by Academy standards, and its downplaying of the possible risks presented by biotech products. Kelman was stunned by the reaction to his committee's effort, saying:

"It's not a perfect document and was not intended to be the end-all. It was a committee report that went through 55 reviewers. What I resent is that some people said that we made an unequivocal statement that biotechnology was safe... We never said that. We simply said that the techniques themselves are safe. And that's true."

Criticisms aside, the Kelman report did not accomplish one of its main goals: shifting the focus of release debates away from the process of biotech to the products and to the environments in which they are released.

Process Phobia Lingers

The next attempt to refocus release debates on products came out 17 months after the Kelman report. In February 1989, the Ecological Society of America (ESA) issued another report on the risks of releasing genetically engineered organisms into the environment. Because it was both written and reviewed by many prominent ecologists and because of its formal, scholarly appearance, the ESA report quickly became everybody's favorite document on releases. Ironically, the chairman of the committee that authored the ESA report, Dr. James Tiedje from Michigan State University, is a good friend and colleague of Dr. Kelman, and his committee said virtually the same thing as Kelman's about the process of biotech:

"Genetically engineered organisms should be evaluated and regulated according to their biological properties (phenotypes) rather than the genetic techniques used to produce them."

The key difference between the ESA report — "The Planned Introduction of Genetically Engineered Organisms: Ecological Considerations and Recommendations" — and the Kelman report is that the ESA report accepts the fact that profoundly new processes will beget more scrutiny. Therefore, a certain amount of process focus — and phobia — will remain for some time to come. The fact is, the biotech toolbox includes techniques that can produce unique products impossible to imagine just a few years ago. (See *What Is Biotech?* on page 168.)

The Kelman report said:

"There is no evidence that ***unique hazards*** *exist either in the use of R-DNA techniques or in the transfer of genes between unrelated organisms. The risks associated with the introduction of R-DNA-engineered organisms are the* ***same in kind*** *as those associated with the introduction into the environment of unmodified organisms and organisms modified by other genetic techniques." [emphasis added]*

The Kelman report said that even though some biotech processes and products are new to the world, they do not present new risks. That's true, said the ESA document, but the "newness" biotech brings, by definition, means it will be scrutinized more closely. The ESA report said:

"Because many novel combinations of properties can be achieved only by molecular and cellular techniques, products of these techniques may often be subjected to greater scrutiny than the products of traditional techniques."

In other words, biotech can beget organisms never before seen under the sun and when it does, the risks it brings with it may be minimal, but nonetheless new.

Keep in mind that biotech is no different than any other genetic technology until it is used to do what it can and the other genetic technologies cannot — namely catapult genes over long-standing evolutionary barriers.

Limitless Leaps

Biotech was born when a toad's gene was spliced into a bacterium's. Now clearly, toads and single cells don't mate in nature — heck, they don't even date. Even this early use of biotech showed that it has added the possibility of limitless leaps in the evolutionary process. *Genes from any organism can be spliced into or mixed with those from any other organism.*

Toads and microbes don't mate in nature — heck, they don't even date.

Biotech thus provides humans with the power to add to the tree of life organisms that Nature never envisioned.

To be sure, Mother Nature is still the genetic-engineering champ — a title she has owned for billions of years. Biotech is often based on the understanding and application of her tricks. In fact, the surest way to frustrate a normally unflappable scientist is to forget this fact and suggest that rDNA is a recent addition to the world. French microbiologist Dr. David Tepfer adds important perspective here:

"Recombination is a process that takes place in almost all genetic systems. There's nothing special about rDNA; almost all DNA is the product of recombination. For anyone to say that Frostban was the first release of an rDNA microbe is more than unfair to the scientific community; to pretend that recombination hasn't been going on all the time is dangerous and wrong. It suggests that scientists think these sorts of microorganisms are extremely dangerous and that just doesn't reflect reality.

"You have to remind people, even scientists, that genetic transformation and horizontal exchange is a perfectly natural process that's been happening since time zero probably. And that this transfer of genes, from bacteria to bacteria or bacteria to plants, is a perfectly natural thing that's been taking place since

the earliest moments in evolution. It's like plant breeding — a very ancient process. What's being done with biotech is in this way part of a continuum — exploiting natural systems and accelerating and expanding them; getting organisms to accept genetic information from other organisms, even from other kingdoms."

Debates of the 1990s

While the 1980s will be looked back on as the decade of debate about the processes of biotech, the debate of the 1990s will shift to the nature of the organisms proposed for release; to the nature of the active and invisible Bacterial Bazaar; and to what happens when you combine these two.

One case perhaps more than any other best illustrates that this important shift has begun. So let's move with it from the debates of the '80s to a foreshadowing of the showdowns of the '90s.

Case Study: Crop Genetics International

This chapter is a case study of a genetically engineered microbe developed and released by Crop Genetics International, a biotech company in Hanover, Md. This case was chosen because it provides a perfect bridge from the 1980s debates about deliberate releases around the world — with their intense focus on the process of biotech and small-scale field tests — to the debates that will define the 1990s: debates over which genetically engineered microbes the world will deem safe to release into the environment on large scales. And, just as important in the long run, which altered microbes will actually work outside.

Our story begins when a pioneer in plant molecular biology wondered...

"What If..."

In the early 1980s, when "agricultural" and "biotechnology" were words being joined for the first time, pioneering plant biologist Dr. Peter S. Carlson saw the one huge hurdle in the way of genetically engineered crop plants, especially the biggies — corn, wheat, rice, soybeans and other Top 10 crops. They don't respond to tissue culture.

That means you can't take cells from corn and the others, bathe them with all their favorite nutrients and coax them in a petri dish to produce whole plants. You can't culture them, as scientists say, at least not without great difficulty. So what? Well, all the fancy tricks of biotech that enable you to get foreign genes of interest into the cells of these major crops don't matter unless and until you can grow whole plants back from those cells. Why don't these major crops respond as well to tissue culture? No one yet knows.

But what if, Carlson wondered, instead of putting a pesticide-producing gene directly into corn *cells,* you put it into the genes of something that can live inside corn *plants?* Find an organism that can live inside corn plants — or wheat or rice — and you're on your way to a hot biotech product.

Are there such beasts?

A FLOOD OF FIRSTS

Why is the Crop Genetics International case, told in the next few pages, worth telling? Because it represents many "firsts" — some historic and others just plain interesting. But all add up to a rich way to look at lessons learned — and issues that remain — along the long road that the first genetically engineered microbes traveled to see the light of day.

CGI Case Marks the First Time...

...a genetically engineered microbe is developed to combat a major pest of a major worldwide crop: corn

...a biotech company develops a system to put biopesticides into bacteria that live inside plants

...some of Washington, D.C.'s biggest names form a Social Responsibility Committee to help guide a biotech firm through the unpredictable process of winning approval to release its first microbes

...a genetically engineered microbe is released by an agency of the U.S. government: the Agricultural Research Service (ARS) in Beltsville, Md.

...a biotech firm plans to simultaneously release a genetically engineered microbe in the United States and in a foreign country — in a field near Montfavet, in the south of France

...the deliberate release of a genetically engineered microbe is endorsed, albeit with qualifications, by three leading U.S. environmental groups

...with reporters from around the world watching, 2,000 corn plants are individually injected with genetically engineered microbes

...for unclear reasons, the French government says "non" to a deliberate release

Yes. They're called endophytes: bacteria that can live *inside* plants.

Like most plants, corn comes complete with its own endophytes, but they're poor candidates for pesticide carriers. First, corn normally hosts only a few kinds of endophytes — and they don't exist in all corn plants uniformly. Moreover, most are "facultative," which means that they can live *outside* plants as well as inside. And the one thing you don't want to create is mobile, pesticide-packin' bacteria that can move around and take up residence just about anywhere.

So Carlson, a colorful character with Indiana Jones spectacles, screened more than 1,000 endophytes and found one that normally lives in Bermuda grass but will happily inhabit corn. It's called Cxc — *Clavibacter xyli* subspecies *cynodontis.* Little is known about Cxc, which, as we've seen, is true about most microbes that live outside. Cxc is known to be "obligative;" that is, it can *only* live inside plants. And it is thought to be, as it is oft described in the press, "innocuous," "harmless" and "benign."

That is, until it gets genetically engineered. Then the adjectives change to "souped up," "futuristic" and "designer mutant."

The "Good Guy" Killer

The difference between Nature's Cxc and Carlson's is that Carlson's contains a gene from another bacterium that, without biotech, would not find its way into Cxc. From *Bacillus thuringiensis* subspecies *kurstaki,* called Bt for short, Carlson got a gene — the Bt delta-endotoxin gene — and spliced it into Cxc.

Unlike Cxc, lots is known about Bt. Discovered in Germany in 1911 in the province of Thuringia where it was killing larvae of the flour moth, Bt was first registered as a biopesticide in France in 1938 and later in the United States in 1961. It has long been a workhorse biopesticide, used worldwide. In 1983 alone, the World Health Organization (WHO) spread more than 500,000 pounds of Bt across West Africa to control disease-carrying blackflies. Hundreds of tons are applied to trees, vegetables and crop plants each year in the United States as well. According to *Organic Gardening* magazine (12/87), Bt "protects forest trees from Alaska to Georgia from defoliation by spruce budworms and gypsy moths, and broccoli and cauliflower from cabbage worms. Farmers use Bt to protect tobacco and cotton from loopers and budworms, and fruit growers use it against several types of leaf rollers."

Why is Bt so popular?

Because its specificity makes it safe. Bt kills only caterpillars and moths from the order Lepidoptera, beetles from Coleoptera, and some species of flies and mosquitoes from Diptera. Amazingly, more than 4,000 strains of Bt have been isolated, each of which has one or more genes that kill certain bugs and only those bugs. The reason: The various genes in different strains of Bt have, over the eons, evolved into highly specific expressors of toxins that kill only certain bugs. The best part is that Bt doesn't harm mammals, birds, fish or even other insects, because their bellies are acidic like yours and mine and Bt toxins only act in alkaline settings.

A drastic diet deterrent, Bt toxin ruptures cells in pests' alkaline stomach walls and paralyzes the muscles, rendering its victims unable to digest food. They stop eating and die, which is the whole point of pesticides.

"As far as pesticides go," Dr. Rebecca Goldburg of the Environmental Defense Fund in New York sums up, "Bt is a good guy."

"Incide" A Biotech Start Up

Not a bad idea to build a company on: Use cutting-edge technology to produce a microbe not yet seen under this sun — and thus patentable — that can live inside major crop plants and pack a proven pesticide even environmentalists like. So Carlson and John B. Henry, a Wall Street lawyer with an entrepreneurial bent, formed Crop Genetics International (CGI) in 1981 to exploit Carlson's endophytic delivery system, cleverly dubbed "InCide.™"

With experience on the U.S. Senate Foreign Relations Committee staff, Henry attracted notables like President Ronald Reagan's arms negotiator Paul Nitze, who invested $500,000 in CGI, and Sol Linowitz, former Xerox chairman and Carter administration senior diplomatic trouble-shooter, who joined CGI's board. Linowitz told *The Washington Post* (3/21/88):

"I don't go on many boards, but this is one that appealed to me, because if it succeeds it will have a dramatic impact and be a major contributor to the problem of providing biological crop protection. I used to serve as President

Carter's chairman of the Commission on World Hunger and had some sense of the problems to which Crop Genetics is addressing itself. I found myself more and more interested in what they were trying to do."

Henry shrewdly used his Capitol credentials and contacts early — and would use them again.

Borer Basics

CGI's first business target is a chunk of the $350 million annual market for chemicals used to combat corn pests in Europe and the United States. First in CGI's sights is the European corn borer, a compulsive eater that each year dines and dashes on more than 70 million acres across the United States, chewing a hefty $500 million out of the U.S. corn crop. What's more, the EC borer is an international corn terrorist: It destroys more European corn than any other maize menace and is guilty of turning maize surpluses into major messes in Kenya and other parts of Africa.

First introduced into the United States in 1908, the borer is so called because when its larvae hatch from eggs that female moths lay on corn plants, they bore into those plants and eat out the stems. This weakens the plants and they fall — "lodge," as farmers say — or their ears fall to the ground out of reach of mechanical harvesters. And once inside corn plants — or any of the 200 plant species borers bedevil — borers are literally shielded from chemicals by their victims. It's easy to understand why, even though U.S. farmers spray more than $50 million worth of toxic pesticides every year trying to block borers, they only nail about half of them. On top of this, up to three generations of borers can chew through one corn crop, which means more than one spraying — each heavier and heavier as the corn plants grow bigger and bigger and add more and more leaves that shield borer eggs.

In this dim light the theoretical beauty of "InCide" shines through. The scheme goes like this: When the borer bites the corn plants that contain CGI's genetically engineered endophyte, the borer gets a mouthful of Bt-filled endophytes and bites the dust because of a fatal bellyache. What's more, the endophyte's alive and reproducing; so these microscopic pesticide factories are already "in-stalk" when subsequent broods of borers are born to boring lives. Bad news, borers.

What the people who get paid to watch biotech — regulators, that is — are concerned about is that the engineered microbes will move to other places they should not go. But Cxc is not known to live long *outside* plants, so the beneficial bacteria perish with the harvested plants. Thus, the treated plant becomes a containment facility of sorts for the bacteria, making regulators more comfortable that the bug won't spread and make neighboring weeds, for instance, resistant to the pests that feed on them.

That CGI's microbes die with the crop when it's harvested also means repeat customers for CGI, because growers must come back every year to buy new endophytically treated seeds — just as corn farmers now buy new hybrid seeds each year. Nature's own biological barriers prevent the reuse of either product.

Henry confidently predicted in *Business Week* (4/18/88) that InCide "will be a $100 million product...There'll be 1,000 plant vaccines in the next 50 years. Ours will be the first."

Surely a scenario for success — or is it?

Performance Time

Good ideas don't kill pests, products do. And, by the spring of 1986, pressure was mounting to test InCide outside to see if it works where farmers do.

Having seen the public pillorying that AGS endured with Frostban, Carlson and Henry were determined not to make the same mistakes along the way to winning approval — both regulatory and public — for its first field tests. Carlson clarifies:

"After AGS and others had so much trouble, it was very clear that we were dealing with a public policy issue, not a science issue. So that's how we approached it."

And that's when John Henry realized it was time to talk to his Capitol contacts again. In April 1986, he approached William D. Ruckelshaus, administrator of EPA under Presidents Richard Nixon and Ronald Reagan, and asked Ruckelshaus to help guide CGI through the public policy process the right way. Ruckelshaus agreed to help because, according to Philip Angell, his top aide, "Bill's very interested in how the regulatory process affects the larger public policy process and vice versa, and biotech presents some *really* interesting public policy issues — it's cutting-edge stuff."

So Ruckelshaus turned to his Rolodex that resonates with big names, called some of his friends and formed CGI's four-man Committee on Social Responsibility. Joining Ruckelshaus were:

Elliot L. Richardson, former secretary of Commerce, former secretary of Health, Education and Welfare, and former secretary of Defense — not to mention, which CGI always did, former U.S. attorney general and ambassador to Great Britain;

Douglas M. Costle, dean of Vermont Law School and also a former EPA administrator;

Robert M. Teeter, perennial Republican pollster and president of Detroit-based Market Opinion Research, a leading polling firm during the last three presidential campaigns.

Big science and big names in place, now it was on to the big time — showing Wall Street and the world that InCide works outside. But first, CGI had to win approval for small-scale field tests.

Telling Time

The big advice from those heavy-hitters on CGI's Social Responsibility Committee: Go tell everyone who is affected by your plans and talk to them openly and honestly. But most important, find out what *they* want to know. It's amazing how long it had taken the biotech industry to seize on so sensibly simple a strategy. Angell wonders:

"How an industry that includes many of the biggest companies with real Washington savvy could stumble for so long on an obviously small 'p' political problem — that being the first into the environment with genetically engineered microbes demands scrupulous openness and honesty — is beyond me."

CGI followed its committee's advice and swung into action. Peter Carlson picks up the story here:

"The committee said: Go in and talk to the regulators and determine what data they consider important, outline your plans and get their reaction before you even do an experiment. Shop the ideas. Make sure that the questions you're asking are questions they want answered. Once you've established the right questions, then generate protocols, so that the way the experiments are done will meet the expectations of those who will read and regulate them.

"Next thing was go out and talk to the politicians. Politicians don't like to be surprised. So we did that in Maryland many months ago to let them know what we were planning on doing.

"Next was public meetings, so we did some of those. And, generally, you can convince the public that scientists have sons and daughters like everyone else, that there's nothing especially unusual about a scientific person, and that oftentimes the Rorschach tests of horror that come to the public mind about biotech really have less to do with fact and more to do with perception caused by hype.

"Next: Go out and talk to environmental groups; don't surprise them either. Get everybody involved. In this situation, where regulations really aren't clear, brief as many individuals and groups as you possibly can, so no one's surprised and everyone knows what's coming. Put all the data out there. And you've become responsible citizens."

Pausing in thought, Carlson takes the kind of slow, big breath that lets you know the bottom line's not far behind:

"My cut on it is, either we can go on spraying chemicals into the environment and poisoning ourselves or we can get out and look for alternatives. Biotech is a credible alternative, but in the absence of field tests and commercial releases, you'll never know how real that credibility actually is. So getting something done in a safe and open manner, above board, and answering the question 'Does this have possibility?' is very important."

Now it was time to sell InCide to the two most important markets for any biotech product. First, the users, which in this case meant corn farmers with borer problems. Second, the public, which in all cases can make or break a product, if not a business, on perception alone.

Test Where the Problems Are

CGI proved early the merits of an old Southern business wheeze: "Don't sell solutions to folks who ain't got problems."

Maryland corn farmers definitely have a problem. Maryland corn has become a borer favorite in recent years, with as much as 20% of an average corn field lost to borers' appetites. Faced with few alternatives, most farmers blitz borers with FMC Corp.'s Furadan, a proven pest-killer.

To be sure, the farmers living close to the proposed test site in Queen Anne's County, Md., are as interested as any in finding alternatives to toxic pesticides. To suggest otherwise is to suggest that farmers aren't concerned about the health and environmental impacts of ag-chemicals. That's just not true. But chemicals work; they kill pests on contact — borers included, that is, when the chemicals actually get to the borers. And in order to be attractive, bioalternatives to chemical solutions have to work just as well. That's reality.

Like many chemical pesticides, however, Furadan kills other living things as well. In fact, because Furadan is known to kill birds — including the American bald eagle — the EPA on January 5, 1989 recommended banning some forms of Furadan.

Now, farmers have long known that the only way to know whether something works is to give it a try. As Queen Anne's County farmer Jack Ashley pointed out to *The Washington Post* (12/13/87): "My gosh, you never find out anything unless you try." And it's not especially surprising that CGI found a receptive audience in borer-infested farm country.

But CGI was testing more than just a product; it was also testing a *process* — the federal regulatory process. So conducting a second, simultaneous field release where those who matter in Washington, D.C. would get a close look made sense.

A Huge Ally: The U.S. Government

While it made sense for CGI to go where farmers are besieged by borers for one field release, it made news when CGI wrapped itself around the U.S. Department of Agriculture for the other release. CGI formed a "cooperative research agreement" with the USDA's Agricultural Research Service to try InCide outside and, in so doing, deftly pushed the whole consideration of deliberate release to the fore in federal regulatory and policy circles.

Henry put it straight to the *Wall Street Journal* (5/10/88):

"Here we have this right arm of government, this Paul Bunyan bicep in its Popeye form that regulates, and we have this deformed left arm in a sling when it comes to promoting positives. So we decided that rather than sit back and wait for ourselves to be regulated we would be more activist in terms of getting the imprimatur of the government to work with us."

The ARS, according to its own pamphlet on the CGI agreement, is the USDA's "major research agency, devoted to solving agricultural problems of national and international scope; ARS has responsibility for developing improved pest control methods and for cooperating with the private sector in these efforts." ARS has been in the borer-busting business since 1950 when it created the corn borer laboratory in Ankeny, Iowa — the same year the borer first invaded the nation's corn belt. However, borers have made steady advances in the 40-year-old borer wars, so ARS welcomed any allies it could find.

Furthermore, Congress perennially wonders aloud what actually goes on out at the ARS's Agricultural Research Center, a sprawling, 7,000-acre facility that would fit in a lot better somewhere in rural Iowa than on the edge of explosive development in Beltsville, Md. As Peter Carlson points out:

"Every year when OMB takes a hard look at government deficits and government assets, it never fails: Someone points out how much the deficit could be reduced by building condos on the 7,000 acres of prime development land that is Beltsville. Then an ag-type has to reason why the land should be kept in corn research instead of condos, and the point is made that ARS does research at Beltsville to, among other things, keep our farmers competitive. We just offered them the opportunity to take part in some cutting-edge stuff."

ARS gladly accepted, chose a Beltsville site for the proposed joint field test and set its sights on the summer of '88 for its first outdoor test of a genetically engineered microbe.

The French Connection

As *A Flood of Firsts* on page 64 indicates, CGI also intended to test its first application of InCide simultaneously in France, where the EC borer is corn farmers' No. 1 nemesis. Actually, CGI had considered conducting the French tests as early as 1987, but was convinced otherwise because, as Philip Angell, then vice president of William D. Ruckleshaus & Associates, reveals:

"It would have put the French in the awkward and potentially embarrassing position of testing a U.S. company's microbe before the U.S. had approved it. More importantly, it would have put EPA and CGI in tough positions because it would look like CGI was trying to end-run U.S. regulations."

By *not* asking the French for permission to release InCide in 1987, CGI avoided for a time the pressure put on all parties when an American company tests its microbe in a foreign country's environment first. But CGI's desire to test Cxc in the French countryside remained clear to all. And French authorities continued to evaluate CGI's data. Would the French give the go-ahead to CGI before EPA had weighed in with its judgment? What would the French do if indeed the EPA delayed or denied the tests on any grounds? The answers surprised many observers.

By the spring of 1988, CGI was ready to test InCide outside in a field near Montfavet, in the south of France. It seemed a foregone conclusion to most observers that the French would say "oui" to the small-scale tests. After all, the French had been very lax about reviewing the release of altered organisms. Dr. David Tepfer, an American who was raised and works in France, predicted in early April 1988:

"I don't think there'll be any reaction to CGI's test in France. The French are not reactive in that way. The French let something go until it gets extremely serious and then instead of protesting they have a revolution. There's no intermediate ground for the French. It's all or nothing."

In fact, until early 1987, the French had no agency or body to review deliberate releases. The Commission Genie Biomoléculaire (Biomolecular Engineering Committee) was created in late 1986 to review releases, but only considered projects involving INRA, France's equivalent of the USDA. The CGB remains the only formal reviewing body for outdoor releases in France; the French don't have any central environmental reviewing agency like the EPA.

Thus, while CGI's proposal was subject to review by INRA's CGB, all safe betters had their money on approval. As Tepfer, whose twin brother Mark sits on the CGB and is also a scientist at INRA, indicated in April 1988:

"The committee is all very low-key and very informal at the moment. It's positively disposed toward the CGI experiment."

On May 10, 1988, John Henry confidently claimed to the *Wall Street Journal:* "...the French are goo-goo on biotech and they love agriculture."

It looked in the summer of 1988 that all systems were go in France for CGI.

Small is Beautiful...

To conduct U.S. field tests of its endophyte, CGI had to obtain regulatory approval from the USDA's Animal and Plant Health Inspection Service (APHIS) and the Environmental Protection Agency. Because both ARS — CGI's testing partner — and APHIS are part of USDA, accusations of conflict of interest arose immediately. Could one agency promote biotechnology by field-testing one of its first major agricultural products and at the same time fairly evaluate whether the test posed significant risks?

Dr. Arnold Foudin, chief biotechnologist at the USDA's Biotechnology Permit Unit, responds:

"We're harder on our own because we can't afford to be the first accident. Can you imagine the fate of someone whose bug takes over Beltsville? Besides, the harder we are the better, because wait till they get over to EPA; that's when they'll be glad we were hard on them."

Foudin's right about EPA. Its evaluation is the one that carries the most weight in Washington when it comes to field releases of microbes (see page 116). Without EPA approval, no genetically engineered microbe gets outside. So the conflict-of-interest chorus didn't have an audience for long and died away.

Starting in December 1987, when CGI applied for EPA's environmental use permit, a sort of license for scientists fishing for results in small-scale field tests, CGI's proposed tests began to be examined under the hottest lights. Written comments came in from the leading environmental organizations already committed to biotech watchdogging — principally the Environmental Defense Fund (EDF), the National Wildlife Federation (NWF) and the National Audubon Society. Some of their comments suggested that CGI's strategy of surprising no one and positioning themselves as environmental allies trying to replace toxic chemicals was selling well. The NWF wrote:

"We commend CGI for attempting to develop alternatives to chemical pesticides. Our dependence on chemical pesticides has resulted in an environmental nightmare. NWF encourages applications of biotechnology that provide environmentally sound alternatives to these pesticides."

The other environmental groups made equally supportive statements. The only group that flatly opposed CGI's proposal was, as expected, the Foundation on Economic Trends — famed anti-biotech evangelist Jeremy Rifkin's Washington-based organization. Able to use the threat of lawsuits as effectively as Luke Skywalker uses his light sword, Rifkin decided in the end to leave his light sword in his belt and not put up a fight to stop CGI, telling *New Scientist* in July 1988:

"The reason we aren't fighting it [CGI's test] is simply that the product doesn't work."

Given that the hallmark of Rifkin's many successful protests against previous experiments had been, as he has said to many audiences, "Once a *living* organism is released, folks, that's it; it can never be called back" — which is indeed a true statement — his tame response didn't make sense. Truth was, Rifkin knew that no court in the land would delay CGI's experiments, because CGI had jumped through all the necessary regulatory hoops. In short, Rifkin had no case against CGI.

In an era when environmental organizations can bring companies and government entities to their knees, their support — albeit *only* for a small-scale test — became a big-scale deal.

...But Big Would Be Bad

Without exception, every one of the environmental groups that submitted comments to EPA clearly articulated a number of major concerns about CGI's InCide that, unless adequately addressed, would cause them to oppose any *larger* scale tests of the endophytes. Concerns such as:

- "Given time and wide-scale use, Cxc/Bt might well be transferred from corn back to Cxc's natural host, Bermuda grass. This is especially true since Bermuda grass is a weed of corn fields and Cxc is spread by farm machinery...Bermuda grass is, however, also a principal weed of sugar cane and cotton fields in this country. Would freeing the grass from some natural enemies increase its weediness?"

  — Environmental Defense Fund

- "Cxc is a relatively unknown, unstudied organism that has only recently been described and named and has been the subject of very little ecological investigation."

  — National Wildlife Federation

Audubon's Maureen Hinkle summed up environmentalists' concerns: "Major unanswered questions stand in the way of wide-scale use of this endophytic bacterium."

What the environmental groups had endorsed was the small-scale testing of a biological product that might replace a toxic chemical — not the large-scale release of a genetically engineered microbe. This opposition to large-scale release put them back on the side of Rifkin, who whipped out his light sword for *Science* magazine on 10/28/88, warning that whoever attempts to test an altered microbe on a commercial scale "will face years and years of battle in the courts and in Congress."

"Chemical Bashing" Backfires

The momentum building behind CGI's experiment-cum-crusade caused it to get carried away with trumpeting the potential of its product and trashing the track record of Furadan — the insecticide it could potentially replace. Just weeks before the EPA was likely to let CGI conduct its first deliberate release, the company inadvertently shot itself in the foot with another kind of deliberate release: a press release. CGI issued a press release that claimed Furadan, the chemical most farmers currently use to kill corn borers, kills 2.4 million birds every year as well — including bald eagles.

In so doing, CGI was like the kid at the beach who gets wound up and kicks sand in the faces of lots of folks he definitely didn't want to offend. Repercussions begin immediately.

Kicking back with much bigger feet, the huge FMC Corp. vociferously complained about CGI's unsubstantiated allegations and forced the deletion of any reference to Furadan in subsequent press releases.

Things got worse.

The members of CGI's Social Responsibility Committee did not learn about the explosive release until *after* it had gone out. Like microbes, once press releases are out they can't be called back. Speedy resignations took place and the committee collapsed. Philip Angell explains why:

"Bill [Ruckelshaus] made clear from the outset that at no time would he tolerate 'chemical bashing' to advance the CGI cause. First, it's no secret that Bill's on Monsanto's board. And even though Monsanto has as its corporate

goal the eventual replacement of its chemical fertilizers and pesticides with bioalternatives, Bill just doesn't believe you build the strength of one product by bashing another. The value of CGI's product, of biotechnology, of the whole approach it brings to nature and dealing with pests is unique — fascinating enough, intriguing enough and ultimately of enough value that it should stand on its own merits. And it can; eventually that will be clear."

After these events, only Elliot Richardson remained involved with CGI, taking a seat on its board.

Fortunately for CGI, the committee's collapse came after its work was basically done: The strategy its members had laid out had been executed with aplomb. And it worked. On May 24, 1988, EPA gave CGI permission to test InCide outside.

The French Say "Non"

Then, to everyone's shock, the French government said "non" to CGI's field test in southern France. What had been viewed by most as a formality, that is INRA waiting for the EPA's decision before saying "oui," quickly became a mystery. No reason was given for the French denial. And, in light of EPA's approval and observations by many top scientists on both sides of the Atlantic, to say the reason was scientific would be subterfuge.

Apparently, something else — something much bigger than one field test — was taken into consideration. Martine Leventer, chief business correspondent for the popular Paris weekly *Le Point,* was as surprised by the denial as anyone. Left to surmise, she simply said, "It must have been due to pressure from outside France — probably from the European Commission in Brussels that's trying to work out rules for release for all EC members. French officials I'm sure didn't want any more trouble — especially after the embarrassment at Dijon." (See page 144 for details.)

The French decision illustrates, on a broad scale, the role national and international politics can play in what appear at first blush to be regional or even local issues. More specifically, it reminds all that science oftentimes has little to do with decisions made on issues that are in fact science-based. Politics, in the end, will usually prevail over science. It's that simple.

Into the Fresh Air

Finally, in early July 1988, at the Beltsville site that included a barren zone around the test plants, an eight-foot security fence, a deer fence, a dike to collect runoff water and round-the-clock security patrols, with the world's media present but without the mayhem that had accompanied earlier field trials, CGI and ARS scientists successfully injected — literally with hypodermic syringes — the "plant vaccine" into 2,200 young corn plants. In a similar scene at the Maryland site, scientists also injected billions of endophytes into threatened corn plants.

Completing what John Henry called the "most expensive agricultural field tests in the history of the world" — about $3.2 million* for roughly three acres of tests — was a sweet victory for CGI. Given what had happened to other attempts to test microbes outside in the 1980s, just getting out the door with one of the first ag-microbes that would not exist *but for biotech* proved a notable achievement.

But once out in the hot light of the sun, InCide, like all biotech products, had to prove its worth and its safety.

Mixed Reviews

The results from the first outdoor tests of InCide were mixed.

First, CGI's Cxc bacteria with Bt on board did not kill corn borers quickly — in fact, they didn't kill many at all. But that's what Carlson and CGI had expected. They knew that they would have to get the additional Bt gene to produce or "express" more toxin in future trials, thus making the Cxc carrying it more potent.

Second, in more apparently bad news, CGI's engineered bacteria also depressed the yield of the corn plants into which they were injected. While interesting, like the low-and-slow kill rate, the yield dampening was also not a surprise to Carlson. Both results were scientific problems that he had expected and would surely have to address.

But how well CGI's — or anyone else's — genetically engineered microbe did what would make it potentially commercial didn't really

*This figure includes all the safety and security precautions mentioned above, as well as the public relations costs for the many press briefings prior to, during and after the actual tests, and the scientific tests conducted to see whether Cxc moved off the test site and how well it performed.

matter in this early stage. To claim otherwise and say that, in CGI's case, because it didn't kill corn borers on contact it failed is to miss the entire point of small-scale field tests. You see, *the first field tests of InCide, like virtually all field tests proposed and/or carried out in the 1980s, were conducted to gather information on how genetically engineered organisms behave in the environment, not to prove that any given organism can perform like a superstar right away — only Bo Jackson can do that.*

The fact is, initial field trials are required and designed to prove, more than anything else, that biotech bugs do not pose unacceptable new risks, if they pose any new risks at all.

Critics' Chief Concern

The biggest concern that remained after the small-scale field tests was the same as the biggest going in: unintentional spread. How much Cxc got to places it wasn't intended to go?

To everyone's surprise, Cxc were found hitching rides on some local beetles — corn flea beetles to be exact. Remember, not only were Cxc not supposed to live outside the insides of plants, they were not supposed to take natural taxis to destinations unknown.

Bad news: Mobile borers.

While this was clearly not good news to CGI, here's how they handled it.

Immediately after the first field data were available, they collected flea beetles in the test sites, then sprayed those sites to kill any remaining beetles. Next they took the captive beetles into their greenhouse and let them pig out on Cxc-loaded plants until the beetles were plump with Cxc. Finally, they put these bacteria-laden beetles on plants with no Cxc in them and let them feed on them for a few months. The question was, did the non-colonized plants now contain the recombinant bacteria?

"And the answer is no," says Carlson. "So the real point is that while you can indeed find Cxc on a few beetles, those bacteria are not likely to colonize other, non-target plants."

Does this answer the spread question? If you are CGI, yes. If you are the regulators, it answers it well enough. If you are any of the environmental groups, as we'll see, no. But these differing responses are standard when science is debated in regulatory circles: It's always a matter of perspective and degree.

The Big Picture

The real worry with "unintentional spread," meaning where engineered organisms will wind up, occurs when releases take place on large scales. After they have been released on thousands, perhaps millions, of acres, where else might biotech microbes set up shop and go into business? In CGI's case, what other plants might Cxc colonize and make better at fending off predators with its lethal Bt gene?

The National Audubon Society said in its initial review of CGI's plans to test Cxc:

"At least 29 potential host species in at least 10 different plant families can act as good or excellent hosts of Cxc. This suggests that Cxc can successfully colonize a range of plant species."

What's more, Audubon and other organizations also pointed out that Cxc might be spread by other insects, on seeds, in run-off water and by mechanical means — such as harvesters, shears and shovels.

So not only do Cxc have quite a few good places to stay, they also have various ways of getting there.

Indeed, CGI fully recognizes that Cxc is not perfect. Nothing is. What's revealing is that CGI chose Cxc as the bug upon which to base an entire endophytic delivery system because its "fate and spread characteristics" were *better* than other endophytes they screened. But "better" is in the eyes of the beholder, because those same fate and spread characteristics are precisely what worry environmentalists about Cxc. Carlson explains:

"CGI looked at hundreds of endophytes and we chose our bug because of its fate and spread characteristics. That is to say, it can't live long outside plants and it had very little spreading characteristics. It does have a number of potential hosts, as most bacteria do. But, even if Cxc is spread in any form to a new host, that Cxc, the engineered version, does not get into the new host's seeds and therefore cannot become a permanent part of that plant's makeup. This is a very important and little understood point: The endophyte cannot get into the seeds of plants or the kernels of the corn in this first case. The plant fluid or 'blood' is not shared directly by both; there is no mixing of 'blood'. And since Cxc only lives in the fluids of plants, it does not get into the seeds. This is true in all species that we have looked at with Cxc.

"What's more, we have found that the Cxc with Bt gene multiplies more slowly than its 'wild type' or parent Cxc; it's like two twins, but one — the engineered version — has extra luggage to carry: Which do you think will get to the top of Mt. Everest first? Now, when you add this to the fact that the engineered version naturally loses its Bt gene over time as it multiplies — in about every 1,000 divisions, the Bt gene gets chewed up by enzymes in the Cxc — you realize that even if Cxc/Bt did get moved and did exchange genes, the product would eventually die out.

"Now that's the package that nature put Cxc in and it's the best option of the ones screened so far. We realize that it's imperfect, but we also know that many farmers and more consumers are fed up with chemicals."

CGI has obviously done considerable homework on Cxc and environmental groups acknowledge this, as this comment by the NWF indicates:

"We commend CGI for the large quantity of natural history data the company has generated on Cxc and Cxc/Bt the past two years."

But does CGI's extensive work on understanding how Cxc behaves in nature put to rest the key environmental concerns brought to light by the first field trials? Not a chance.

Once again, it comes down to different perspectives on the assessment of risk.

Prelude to a Showdown

The irony inherent in all the time, money and media spent on the small-scale field tests of InCide is that no one had much concern about testing Cxc on small scales. What everyone who cared to comment cared about the most are large-scale field trials of Cxc — and any other genetically engineered microbes.

So when, in December 1988, CGI asked the same federal agencies for permission to conduct the second round of field tests of InCide on a total of 12 acres on eight sites in four states, the comments that came into the agencies were virtually the same as those for the original tests: Go ahead and test on these small scales, but think again when you prepare to go to large scales.

Dr. Goldburg from the Environmental Defense Fund said it most succinctly in her written comments to the EPA:

"Despite the relative safety of CGI's proposed experiments, we see little prospect that Cxc/Bt can ever be ecologically acceptable for commercial pest control."

The National Wildlife Federation seemed to look past CGI's second round of tests, eager to get in the ring for the title bout:

"CGI will soon apply to EPA for permission to conduct larger-scale tests. Because this application will raise many significant new issues, we urge EPA to look beyond the small-scale tests to the rapidly approaching issue of evaluating large-scale tests and commercial uses of novel microbial pesticides."

So what is large scale? "That's a great question," Carlson says with nervous laughter, "and nobody seems to want to answer it." Looking ahead to the early 1990s, Carlson admits that:

"Environmental groups and people like Jeremy Rifkin will define the place where and the time when the battle over bigger tests will occur. Most folks think we're crazy to continue, but somebody's got to do this because it moves the U.S. and the world along in the debates on the merits, based on a lot more data, and gets us away from fear and hype. While all of this moves everybody involved down the pike toward confrontation, in a sense it also holds off confrontation, because we're getting information that makes it harder to confront us. It becomes harder to whip up excitement about this."

Throughout the summer of '89, CGI successfully conducted its second-round field tests, each of which cost "around $250,000," according to Henry, "which is ridiculous and cannot go on for long." Very little was written or said about these second tests, because, again, they had yet to reach scales that mattered to anyone but CGI — and its investors.

What About the French?

While the second small-scale tests were being debated and conducted in the United States, what were the French doing about testing InCide outside?

Nothing but jawboning, really.

The French decided to postpone any outdoor release of InCide, voicing two concerns — one on the record and one off. On the record, French officials told CGI and anyone who asked that they wanted more data on the proposed tests under French conditions; they said that the U.S. data on spread did not apply to France. CGI, in cooperation with INRA, is working on delivering that data.

Off the record, French officials admit that the European Community is embroiled in a huge debate about deliberate releases that extends far beyond France. Many Germans and specifically most German Greens are dead set against releases on any scale. (The European scene is discussed in more detail in Chapter 7, starting on page 135.) So don't hold your breath waiting for a French release of CGI's Cxc.

CGI will, as Carlson says, "continue working with INRA, which has many good scientists," and proceed slowly with caution. Carlson says bluntly:

"There's no sense chewing up cash in someone else's political fight. So we will develop data and hopefully be first in line when they're ready to test."

What's Ahead?

The next step for CGI is to test a more souped-up or "efficacious" model of their Cxc — namely one that produces more Bt toxin to kill more borers — outside. This CGI will do on its test site in Maryland in 1990. This single-site, small-scale test will not force the issue of large-scale release. That CGI will likely do in 1991.

As for France, no tests are planned for the near future.

Interestingly, CGI also tested its Cxc system, with a different Bt gene, on rice plants in Maryland in 1989. As Carlson points out with clear pleasure: "CGI had the only rice paddy growing in the state of Maryland in 1989." Now, when you think of rice, you naturally think of Japan. But, for reasons that aren't at all clear, the Japanese are very reluctant to allow any genetically engineered microbes outside — especially those imported by a U.S. firm (see page 137). Thus, CGI will not be going outside with Cxc in Japan anytime soon.

As Carlson says, "CGI is a small North American company focusing first on big North American problems." When will CGI — or anyone else — get a chance to try and solve any big agricultural problem with a genetically engineered microbe? That remains a wide-open question not only in the United States, but around the world as well.

Risk: The Bottom Line

Back in Chapter 2, you learned that the 1990s will be remembered in biological history as the decade of debates about two questions: Are genetically engineered microbes safe to release into the environment on large scales? Do they work? This case has carried us into the 1990s by posing precisely these two questions.

To answer the second question you must first answer the first — and to do that you must test genetically engineered microbes outside *on large scales.* That means deciding who's taking the bigger risk: the companies betting that their microbes will become safe and effective replacements for toxic pesticides and harsh fertilizers, to name two examples, or environmentalists who seek to delay products because not enough data are in to ascertain the safety of releasing biological alternatives on large scales.

Clearly, we need to take a closer look at different perspectives on risk.

5 Wrestling with Risk

"This is the only country where 400,000 people die every year because of tobacco and, what do we do? We ban artificial sweeteners...because a rat died!"

Comedian George Carlin on his album
"How Did I Wind Up in New Jersey?"

"How extraordinary. The richest, longest-lived, best-protected, most resourceful civilization, with the highest degree of insight into its own technology, is on its way to becoming the most frightened. Has there ever been, one wonders, a society that produced more uncertainty more often about everyday life?... Chicken Little is alive and well in America."

Aaron Wildavsky, co-author of *Risk and Culture,*
quoted in *San Francisco Chronicle,* 6/1/86

"In the absence of understanding, we 'shy at kittens, and cuddle tigers.'"

"The Public Understanding of Science,"
The Royal Society, London, 1985

Perceived Risk Rules

Ever have one of those frustrating, get-nowhere dreams like, say, trying to pull yourself out of a pit full of snakes on a rope made of lime Jell-O? Trying to explain public reactions to various risks can quickly become a similarly maddening exercise. Fact is, much of the public's reaction to risk is just plain inexplicable. Consider the following, for example.

Which do you worry about more: riding in a car or flying on a DC-10 jumbo jet? Auto accidents are the No. 1 cause of injuries in the United States — 2.6 million every year. But the risk per mile of being seriously injured in a commercial airplane — arguably the safest form of travel per mile ever devised — is about 10 times less than the risk of going by car. How about this: More Americans drowned in their bathtubs in 1985 than were killed in terrorist attacks. Yet thousands of tourists canceled their European vacations in 1986 for fear of terrorists; how many do you think canceled their baths for fear of tubs?

When dealing with *familiar* activities, all of us make our own decisions and seem to ignore the risks. We don't really care that experts tell us that sitting in the sun, drinking, smoking, driving too fast and even climbing stairs or taking baths are all risky businesses. We think we know what we're up against, so we take the risks and get on with life.

On the other hand, when we face *unfamiliar* risk-producers — especially when they involve technology that we don't understand, like biotech — we turn to the experts for advice. What else can we do?

But no matter what the experts say, our perception of risk can be as important as real risk — oftentimes more so. If you don't think so, just ask those pioneering developers of Frostban; they know full well the potency of perceived risk (see page 51).

Myth Busters

Any discussion of risk can be seen as a big red onion. What we've just been through — the quirky, go-figure way the public decides which risks to worry about — is the tough outside skin. You can't miss it, you could chew on it forever, but you simply have to peel it back to get at the juicier parts of the risk story. Actually, a couple more layers of the risk onion must be peeled back and dumped right away. They represent two myths that prevent progress toward the heart of any risk discussion.

Myth #1: *There is such a thing as zero risk.*

Zero risk is pure fantasy. No activity can be undertaken without any risk whatsoever.

In fact, as Irwin Remson, professor of applied earth sciences at Stanford University, told *The Stanford Magazine* (Winter 1988), "Insisting on no risk is the greatest risk of all. It becomes counterproductive for society to pursue zero risk because the money to do so is taken out of cancer research or other areas. That diversion will kill people too."

Still, something inside all of us yearns to be told that certain things are risk-free. Sam Karas, then chairman of the Monterey County Board of Supervisors, felt this force when the original proposal to release Frostban was in front of his board; he even said so in *Smithsonian* magazine (8/87):

"I was 95 percent convinced that there would be no harmful damage to the environment, maybe 99 percent. The question came up about that one percent. There are no guarantees in life, but I wanted one and the scientists wouldn't guarantee that it would be 100 percent safe."

Though it's difficult to do, most people can be convinced that there is no such thing as a risk-free lunch. But the very next thing they ask is, "What are the chances of something going wrong? How much risk do I face?"

Myth #2: *Risk is predictable and quantifiable.*

No one can predict or quantify risks with any degree of certainty.

The answer to the question "Is it safe?" always comes down to probabilities. You've heard the jargon of probabilistic risk. "I think, to the best of my current knowledge, that this is safe," or "The risk of this causing harm is one in 100,000." The National Academy of Sciences once estimated the number of U.S. cases of bladder cancer likely in a 79-year period caused by the continued use of saccharine at anywhere from 0.22 to 1,144,000. If that kind of estimate isn't especially satisfying to experts, it's downright unnerving to the public.

Again, a few comments from some non-scientists involved in the Frostban case illustrate this point. David Porter Misso, a farm laborer in Monterey, complained:

"Having read all the [reports]... I find words like 'should,' 'expected them,' 'no foreseeable risk,' 'infinitesimal,' 'probably is very low,' 'not much likelihood,' 'however is low,' 'should not,' 'doubtful,' 'slight'... You know, I'm not a scientist, but those words don't say yes or no. I thought the whole reason for that impact report was to come up with yeses and noes. We didn't get any yeses or noes." (*Smithsonian,* August 1987)

It's no wonder that most discussions about risk never make it past this point. People want what experts can never give, namely guarantees of safety. Furthermore, as University of Minnesota ecologist Phil Regal notes:

"Americans do respect expertise but they really burn at the idea of a technocracy in which the 'experts' make decisions for society without the input of laypeople."

Thus nonexperts remain exasperated and experts remain unpopular. At some point, though, some experts must be trusted and their advice acted upon. Otherwise hibernation becomes the only alternative — and even that has risks.

A Big Point-Getter

The Tote Board on the next page provides a quick way to predict public response to deliberate releases of microorganisms into the environment. Call it the "Tally-the-Trauma Tote Board," a 10-step test that lets you find out fast whether an organism will put the hairs up on most people's necks.

Now have a go: Take a genetically engineered microbe down this chart and check which side of the table it falls on. If it falls on the far right, award it 10 points; on the far left, give it zero; and anywhere in between, you be the judge. The closer the score is to 100, the more likely it is to push all those buttons in the public's mind that caused the angst avalanche that buried the developers of Frostban, as we saw in Chapter 3.

It doesn't take a Ph.D. to see that many genetically engineered microbes will score big on the Trauma Tote Board.

TALLY-THE-TRAUMA TOTE BOARD

Question	*Decreases Anxiety*	*Increases Anxiety*
who made it	nature	humans
who benefits	you	others
importance of benefit	compelling	vague
familiarity of risk	familiar	unfamiliar
arrival of effects	immediate	delayed
seriousness of effects	ordinary	dramatic
amount of control	controllable	uncontrollable
location of risk	indoors	*outdoors*
visibility of risk	visible	*invisible*
object of risk	non-living	*living*

Many altered microbes score high on the Trauma Tote Board.

Knowing the Risks

So what are the *environmental* risks presented by genetically engineered microbes? To answer that question puts you right back at the cutting edge of what's currently known or, more accurately, what isn't known, about how microbes behave outside. Microbial ecology, as we've seen throughout this *BriefBook,* is still a nascent science.

And if precious little is known about how microbes act outdoors, even less is known about how *biotech-generated* microbes act out there. For only a handful have been released around the world so far, and only on a minute scale, ecologically speaking. Data with breadth and depth, the kind data bases are built on and decisions based on, can only come from large-scale releases, which is the Catch-22 of the deliberate-release debate.

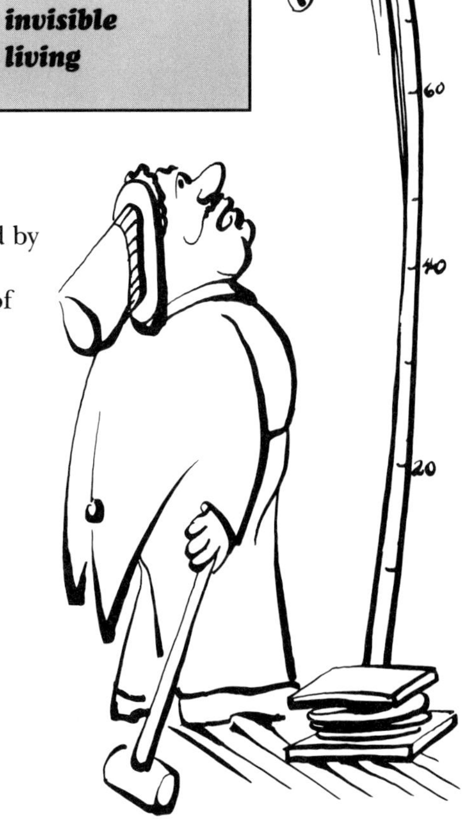

So what's large scale? This question will command the lion's share of the debate about deliberate releases as they continue to expand in scope and number during the 1990s. But we are ahead of ourselves, because scale is a regulatory question and therefore will be discussed in the next chapter.

Right now, it's important to look at the unique risks genetically engineered microbes pose to the environment *on any scale.*

After all, as Winston Brill correctly cautions, "It's not the number of organisms that counts; what matters is whether that organism has an advantage to get the available food over other organisms that are around. If it does, then one cell is as damaging as a million cells." This is because microbes, especially bacteria, can multiply so rapidly, doubling their numbers in less than one hour under ideal conditions.

From Process to Product

The first big point that falls out from the inner layers of our risk onion is that biotech, the process, is really not central to the discussion of *environmental* risks. The method used to alter a microbe doesn't necessarily say anything about how that microbe will behave outside. That a microbe is made different because of biotech — or chemicals, or gamma radiation or UV light or any other mutagenic force — doesn't mean it's more risky or less risky.

With Frostban, the public perception was that biotech had made the release of the ice minus bacteria riskier. No matter how many times it was pointed out that Nature has been making virtually identical bacteria for eons, and that scientists have also made — and released — ice minus using other techniques, it didn't matter. The *genetically engineered* part bugged people.

In response, many scientists eager to try their modified bugs outside have argued that biotech is actually a safer way to alter microbes because of its elegant specificity (see page 168). But biotech's specificity only tells you exactly what alterations you have made in a given microbe. It does not tell you how those changes will alter the behavior of that bug when you unleash it into the environment. So whether you have added a gene, deleted a gene or even moved a gene from one position to another, you cannot say with precision how any change will affect the organism's behavior in a given environment.

Still, it makes sense that if you make a change with biotech that has been made before with other techniques, then there's nothing really new about it and the risks it poses remain the same.

Whether you dig a big hole with a pencil, a shovel or a tractor doesn't make the hole any more or less risky.

Not How, But What

You reach the heart of the risk debate when you focus on exactly what you're putting into the environment—regardless of how it's altered.

Every living thing, including microbes, has a unique set of genes or a "genotype" that describes the type of individual toting it about. Imagine your favorite environment filled with various organisms and creatures; then think of that same environment as lots of genotypes sharing space. The pertinent question becomes: What different genotype is being added to a given environment when an altered microbe is released into it, and what kind of risk does it add to that environment?

It's when biotech is used to leap Nature's barriers and thus put genes into organisms that could otherwise never get there that we step into new areas of potential risk.

What on Earth?

Because biotech has become an "it," it has taken on almost mythical proportions. It can make "brave new organisms," "create life," and even "give us the power of God." But when "it" is taken off its pedestal and brought back down to Earth, biotech may be seen in proper perspective as a powerful series of techniques. Not an "it" but a "they," as explained on page 168.

They — certain biotech tools — do enable scientists to leap barriers that Nature erected and has observed for ages. Biotech gives scientists the power to move genes from any organism into any other organism — frog genes into bacteria, firefly genes into plants, cow genes into fish.

This ability to add organisms to the environment never before seen on Earth — unique genotypes that Mother Nature wouldn't recognize — is the wild card that biotech adds to the risk debate. As a highly regarded 1989 report on risks posed by altered organisms, released by the Ecological Society of America, said:

"Organisms with novel combinations of traits are more likely to play novel ecological roles..."

National Wildlife Federation's Margaret Mellon goes further in a controversial 1988 NWF report, "Biotechnology and the Environment," warning that biotech-generated organisms *will* add risk to the environment:

"At the present time, the most significant risks of biotechnology are to the environment. Engineered organisms can be pests... disrupt the functioning of ecosystems, reduce biological diversity, alter the composition of species, and even threaten the extinction of various species and change climate patterns.

"Unfortunately, such ecological risks of releasing new organisms will **not** *be wholly avoidable.* ***All*** *living engineered organisms released into the environment pose some degree of risk of unexpectedly inhabiting ecosystems that favor their proliferation and that may allow them to do harm."*

Naturally, the next question is, what risks — or benefits — do novel organisms add to the environment *before* they get released?

Many Bad Analogies...

Throughout the early 1980s, the press was for the most part easy prey for alarming analogies to microbial releases. Tales of Kudzu plants (once called "the vine that ate the South"), gypsy moths, starlings, killer bees and many more examples of exotic organisms wreaking havoc in new environments were all used as stern warnings of what would happen when genetically engineered microbes ran amok.

It's true that when *whole* organisms are taken from native environments and introduced into non-native ones they can cause problems. For example, most of the 80 major U.S. weed species originated outside the United States. Before 1900, separating weed seeds from crop seeds was a tough task — and, remember, for decades Congress sent millions of packets of free seeds from foreign lands to eager agriculturalists looking for new crops. By 1923, more than 50,000 types of plants had been introduced into the United States by the USDA alone. Along with these plants came 90% of the pests that plague agriculture today — and unimaginable hordes of microbes, most benign, some good and a few bad. And all invisible. Some of the plants, such as crabgrass, dandelions, Johnson grass, water hyacinth, and, yes, the dreaded Kudzu vine, have been earning nasty reputations as expensive pests ever since.

But there's one big problem with all of the above analogies: While they reveal much about the visible macroworld, they tell us little if anything about the invisible microworld outdoors. Mycogen Chairman Jerry Caulder bristles at any comparison of introducing entire organisms to introducing those with one or a few gene changes:

"It's ludicrous. A one-gene change in a cotton plant or a microbe doesn't make them any less a cotton plant or a microbe than losing a feather makes a robin any less a robin. When a new human is born with millions of new genes, it doesn't suddenly fill a niche that was formerly filled by another mammal. It becomes part of the human niche."

University of Connecticut ecologist Rob Colwell makes a pertinent point on this subject:

"Why not use microbe introductions for microbes, plants for plants, and animals for animals? Those are the appropriate analogies. Otherwise you're talking about apples and oranges. That's setting up a straw man. Kudzu and starlings aren't anything like microbes."

...Few Good Ones

When you turn to microbial introductions, three points must be kept in mind.

1. Of the countless trillions of microbes that have hitched rides around the world on seeds in the mail — or even on the envelope containing those seeds or the person carrying the envelope — only a few have ever come to anyone's attention. Most remain undetected, silent genotypes mixed in with all the other unknown, undetected microbes in whatever environments they wind up in. Dr. Regal adds:

"It would be hard to find a novel, natural microbe that you could say for sure had never blown into your back yard hundreds or thousands of times before."

2. Almost all of the few microbes that do cause problems get noticed because they are pathogens — organisms that cause harm to other organisms, disease mainly. One famous example of a microbial pathogen that's gotten around, unfortunately, is the fungus *Endothia parasitica.* Better known as chestnut blight, it immigrated to the United States on trees sent in 1904 from northern China to the New York Botanical Garden. Before this new version of an old fungus arrived, the American chestnut was the most common tree in Eastern U.S. forests; now it's nearly impossible to find.

3. Examples of non-pathogenic microbes that cause problems are *very* hard to find. One is *Bradyrhizobium* serotype 123, a strain developed and released by a USDA scientist. Says Colwell: "It's a pest in that you can't get other, better, *Rhizobia* established. In some situations it gives higher soybean yields, but in some it's a net drain on the soybeans. It's not a pathogen. It's a pest." And a nasty one, according to Professor David Sands of the Plant Pathology Department at Montana State University:

"Rhizobium 123 is a bad-ass organism, very aggressive as far as colonization, but not a good nodulator or nitrogen fixer. It makes it hard to do new introductions of better Rhizobia strains. We wish we had containment of that organism."

Consider this: Since the late 1800s, hundreds of thousands of "better" strains of *Rhizobia* have been released around the world, yet only a scant few have become rogues. That says far more about what didn't happen than what did.

What *Didn't* Happen

What's most significant about all these *Rhizobia* releases is that they represent perhaps the best example of uncontrolled, unregulated deliberate releases of unimaginable kinds and numbers of unknown microbes into uncountable environments across the Earth — in sum, a worst-case scenario for deliberate releases on a very large scale. For along with the *Rhizobia* went plenty of other invisible creatures, as Professor John E. Beringer, well-known microbiology expert from the University of Bristol, told an international gathering in Cardiff, Wales, in April 1988:

"One of the interesting and very important aspects about Rhizobium inoculants is that almost invariably they contain...at least 10^9 ***unknown*** *microorganisms per kilogram. Very often, where quality control is very poor, the level is nearer 10^{12}, with a few Rhizobia thrown in for good luck.*

"So we have, over about 100 years, enormous experience of introducing a bacterium called Rhizobium which we know something about, plus extremely large numbers of other organisms that happen to grow in the broth culture or in the carrier material used for distributing Rhizobium. We have no idea what most of these microorganisms are, what their properties are, what their potential to cause harm is. But they've been distributed for all this time. Very interestingly, when you look at the regulations that have grown up around the use of Rhizobium inoculants, almost the sole emphasis is on how to look after the Rhizobium and the inoculant...Nowhere have I been able to find any comment...whatsoever that with the handling of such large numbers of unknown microorganisms there has ever been any human health problem or any environmental problem associated with this practice." (emphasis added)

What's most revealing about all these *Rhizobia* releases is *what didn't happen:* None of these countless released microbes made humans aware of them by causing problems. This is probably because "the indigenous guy always wins the war" — a statement Beringer made based largely on his exhaustive study of *Rhizobia* releases across the globe.

Local bugs beating out foreigners brings us to perhaps the most important point about the risks of releasing microbes.

They're Everywhere

Many microbes are ubiquitous. Thus, microbial pathogens probably find their way to virtually every place around the world — unlike most noxious visible invaders, like weeds and insects, whose ranges and mobility are more limited. In fact, 40% of the 155 major microbial pathogens of vegetables have worldwide distribution.

Winston Brill is emphatic on this point:

"Microbes move by wind, by rain, by animals, insects, birds, you name it. Say I've just made a potful of a new microorganism. It's only average in its ability to withstand desiccation, sunlight, etc. And I dump it out in the back yard. That microorganism will go to every single environment on this planet. I can say that categorically."

Indigenous microbes form a tough local crowd that doesn't welcome outsiders.

Then why haven't we heard more about bad foreign microbes taking over local scenes? Because indigenous microbes form a tough local crowd that doesn't welcome outsiders. As Dr. Kelman recognizes:

"I do believe it's much more difficult to shift the soil populations of microbes [than the populations of any plant or animal] and that stability must explain why all these exotic bacteria that are being introduced all the time don't take over. The immediate reaction people have to my comment is how do you know that there haven't been any adverse effects? How do you know that we haven't had an adverse effect with Rhizobia? All we know is that... we're still growing plants; they haven't disappeared."

Ecologist Colwell responds to this point:

"A lot of microbes do have worldwide distribution. But a lot we don't know anything about: They're local in distribution, cryptic in that we can't find or culture them. We need to know more about them. But Kelman's right in a way. There are no examples — except diseases — of things that cause horrible problems. But no one should argue that there's nothing we can do to [harm] the microbial world. If for no other reason because we understand it so poorly."

Nevertheless, the Ecological Society report, of which Colwell is a principal author, stated: "Because redundancy of function appears to be common in microbial communities, in many cases there would be little concern over microbial species displacement caused by an introduced transgenic organism."

Indoor Tests Tell Little

With few good analogies on which to base decisions about relative risks posed by deliberate releases of genetically altered microbes, the next best thing is actual tests. But testing microbes in a petri dish, a pot of soil in a growth chamber, or even in a greenhouse

— anywhere indoors and out of nature — doesn't really tell you much about what's true outside. Brill elaborates:

"The gap between even the greenhouse, the last step before field trials, and a product is many, many years because nature changes the environment all the time and can change it suddenly and very dramatically. Whether it's Ph fluctuations, very hot then very chilly, wet then dry, or that the lighting in a greenhouse is really nothing like sunlight — nature is impossible to mimic indoors. Doing studies on microbes in pots of soil in greenhouses is like conducting psychological studies of people stuffed into closets."

"Doing studies on microbes in pots of soil in greenhouses is like conducting psychological studies of people stuffed into closets."

University of Nebraska plant pathologist Kathleen Keeler simply says:

"Rhizobium in the field and Rhizobium in the lab have nothing in common. Until you try it in the field you don't know anything. About the best you can say now is that microbial ecology is extremely complicated. The relevant questions are only now being asked. Before biotech, we'd never really been able to study microbes outside on their turf, so to speak. Take how they multiply, for example. In a lab a generation time may be as little as 20 minutes or up to a full day, but those are under ideal conditions. In the field nutrients may come and go; when they're gone, it may take up to a year for some microbes to multiply."

And so, according to Dutch microbiology expert Dr. Hans Keunen:

"If you wish to evaluate the chances of an organism's risk in terms of growth rate, looking at a batch culture in a laboratory is very wrong. When would you think that an organism would ever grow at maximal rate, in excessive nutrients in nature? Where? First of all, nature is a very heterogeneous system, that is for sure. Growth in soil will be much more patchy than you would think; nutrients will not be distributed evenly. You will find pockets of nutrients, near a dying plant root, say. Then there will be a whole stretch of no man's land and other pockets which may have an entirely different microbial composition. And then all of a sudden this incredible machinery of a worm comes through and messes everything up and then you can start all over again.

"These things are really never considered by the people who work in the

Worms look like colossal freight trains to microbes.

laboratory; most geneticists simply have never been trained to think in terms that mean the most outside."

Dr. Keeler adds one last, extremely vital point that further reveals why indoor tests to determine what microbes do outdoors don't tell us much:

"Microbes also die. Microbes must ***compete*** *for food and space. Chemicals, on the other hand, don't compete for anything. They work or they don't; it's that simple. And microbial competition is not something we know a whole lot about. Indeed, it's not something we know even a little about."*

Life After 'Death'

Nor is much known about the way microbes die. Except that they do have some pretty crafty ways to hang around for a long, long time without anyone knowing it.

Microbes die, but not, in some cases, quite as finally as you might think. Often, as Dr. James Tiedje, professor in the Department of Crop and Soil Sciences at Michigan State University in Lansing, notes, they don't "die out," they "die back." And there's a big difference. Dying back simply means that they die out below current detection methods, but not off the map, so to speak.

What's more, many microbes can, when times are really tough, produce spores, greatly simplified versions of themselves that slip into a kind of suspended animation in soils. Spores, or endospores as bacterial spores are called, are nearly indestructible and lie dormant for months or even decades, then literally spring back to life as if nothing had happened when the surrounds suit their kind once again.

Pioneer Hi-Bred's Dr. Marshall describes "die-backs:"

"Since microbes exist by the billions, they can lose a billion here and a billion there and it's no more damaging than a haircut. Then, they can go into their Houdini routine and form spores. They appear to 'expire' by dropping below detectable levels, but may have, say, hundreds of copies around just waiting for their day in the sun to return. If I could 'expire' and have hundreds of me around, I'd be perfectly happy.

"We've done studies of soybean roots that amaze me. The day of the first frost everything changes. The microorganisms that were around the roots, you can't find them anymore. A whole new group comes up. They just blossom like crazy. You couldn't find them in July and August. But give them frost and off they go."

Tests That Tell

When you finally arrive at the heart of the environmental risk debate, you realize that the only way to truly learn how genetically engineered microbes — or any other kind of altered microbes — will behave outside is to put them out there and see. That means field tests, on small scales at first. Small-scale tests were to the 1980s what large-scale field tests will be to the 1990s: the hottest topic in biotech's applications outside.

To date, the small-scale field tests of genetically engineered microbes have, more than anything else, supported what the overwhelming majority of experts suspected would happen: The indigenous bugs beat out their much-ballyhooed brethren. Companies' claims notwithstanding, genetically engineered microbes definitely will not dot the shelves of farmers' sheds for some time. No better example of this can be found than what happened to a biotech version of a *Rhizobium meliloti* released on April 19, 1988, in Pepin County, Wis. Having altered the genes and enhanced the bacterium's nitrogen-fixing capability, Biotechnica Agriculture, a Cambridge, Mass.-based biotech firm, hoped to enhance yields of alfalfa by as much as 17%, an improvement Biotechnica had demonstrated in the greenhouse. The test was the first open-air release of *biotech*-altered microbes in the Midwest, and the second in the United States.

So what happened?

Biotechnica's *Rhizobia* got so thoroughly outcompeted by local microbial residents that Biotechnica scientists could barely find them in the test plot soil. Winston Brill explains colorfully:

"Putting altered microorganisms into established soil populations is like dropping a peasant from the People's Republic of China onto the streets of Harlem and saying, 'Go to it, survive and prosper.' He's not going to get anywhere. All these people in New York have been selected for; they chose New York because they can compete and know how to dodge the cars and get the good jobs and be at the right place at the right time. And this guy from the PRC he ain't going to have a chance by himself. And microbes don't have any kind of charities and don't take special care of the newcomer."

Lord, Love a Bug

Even when you ask experts to go out on a limb and come up with the worst-case scenarios for microbial releases, you see blank stares and hesitation more than anything else.

Robert Colwell calls microbes that might go really wrong when they go outdoors "high-risk outliers." He suggests that the average risks from products of traditional breeding and genetic engineering will be about equal, namely zero. But, he cautions, "at the high-risk end, the ability to combine independent evolutionary 'inventions' increases the possibility of creating ecological novelties. These high-risk outliers will arise more often from unanticipated ecological effects than from anything under the biologist's control." This kind of discomfiting, non-specific caution is precisely why microbes are prodigious point-getters on the "Tally-the-Trauma Tote Board" back on page 91.

John Beringer quickly responds when asked about worst-case scenarios:

"We already live with worst cases every day of our lives with the microbes that 'escape' from the treatment of sewage and microbes that might foul our food. There's always a probability, as we know, that salmonella will get into a food process. Occasionally it does. But it's a risk that we take and it's a risk that we live with very, very happily.

"As for soil microbes, well, dragging a plough across a field must be one of the most devastating things that you can do to soil populations and organisms,

and yet it's been totally acceptable since man first pulled a plough. Because we don't see them. You know, if they were large, and we liked the shape and color of certain groups of microorganisms, we wouldn't allow it to happen. If they were butterflies, you wouldn't be allowed to pull a plough across a field, because you would eliminate something that man liked. It's conceptually a different way of looking at a living entity."

Colwell adds:

"We don't care about soil as long as the ecosystem function continues as before. But we should care. We should be clear when we say 'it's more buffered' [than forests or coral reefs] what we mean. I doubt it is. But will it return to equilibrium? That's probably true. Also, the time scale in soil is faster because microbes reproduce faster. We look at these things from a human time scale. I have different ethical feelings about higher organisms than microbes. It's totally bigoted of us."

Environmental Defense Fund biotech expert Rebecca Goldburg believes:

"One of the big questions that nobody is up to facing is, what do we care about natural microbial communities? I mean, we know we care about animals and plants — we can see them; we can appreciate them. But does anyone really care about microbial communities?

"It always comes back to the fact that the microbes that are being modified now affect higher organisms — like Rhizobia with plants and other pesticidal microbes. If we make a mistake it means that we're going to affect organisms we care about. On the other hand, when it comes to modifying microbes to, say, clean up pollutants, what will we be doing to the microbial community? Nobody knows."

Genes on the Loose?

Another potential risk posed by releasing genetically altered microbes is the risk of the new genes themselves persisting in the environment. Remember Bacterial Bazaars — see pages 171-181. It is possible that those creative tricks in bacterial gene transfer will allow particular genes to persist even if the organisms they are initially introduced in die back or out. So says the highly regarded February 1989 Ecological Society of America report on the risks of releasing genetically engineered organisms:

"Transfer of engineered genes from the modified organism to other organisms may occur... through conjugation, transduction, or transformation in microorganisms. If such lateral transfer occurs, an engineered gene may persist in the natural environment even after the genetically engineered organism itself is no longer present."

The ESA report follows the statement above with a long-winded, bet-hedging line that simply says no one knows much about how genes move about in nature:

"The available scientific evidence indicates that lateral transfer among microorganisms in nature is neither so rare that we can ignore its occurrence, nor so common that we can assume that barriers crossed by modern biotechnology are comparable to those constantly crossed in nature."

Startling evidence supporting the image of crafty characters moving genes through Bacterial Bazaars came in mid-1989 — after the ESA report's release — when scientists from the University of Bergen in Norway announced in *Nature* (8/10/89) that they had found viruses in water samples at concentrations *10 million times greater* than previously recorded. One teaspoon of water from Lake Plussee in West Germany alone contained more than *1 billion* distinct viruses. Most microbial ecologists had not focused on viruses' role in ecosystems before this amazing information came to light. But they will now.

Indeed, so many viral taxis, able to ferry genes throughout bacterial communities, is not the most comforting news for anyone interested in releasing new genes in microbes into various environments around the world. As Ian Jones of the Institute of Virology in Oxford told *New Scientist* (8/19/89):

"If a farmer were to spray a crop with a bacterium, for example the ice-minus bacterium..., and the rain washed the bacterium from the leaves into a nearby river or lake, bacteriophages [viruses that infect bacteria] in the water might transfer the ice-minus gene to closely related bacteria in the water."

Even if the ice minus gene were transferred via viral taxis to other bacteria, few scientists feel it would cause problems. But scientists aren't so sure that problems would not arise were the gene getting transferred herbicide or antibiotic resistance — or worse, a gene for, say, cellulose degradation. Ecologist Phil Regal stresses:

"The exact rates and modes of gene transfer are not yet known, thus one can't predict what novelty might be created secondarily in nature if genes move from genetically engineered organisms to other strains. There are some mysterious rules here that we need to understand better. That means we have to find out what the rules are, invent 'auto-destruct' systems or 'leashes' to contain genes, and/or only work with genes that are highly unlikely to cause problems no matter what strains they move into."

Brill readily admits:

"Things that are related can and do exchange material. As species at least. As genera, frequently. There are some genera that are more related than others. And there's indirect evidence that everything has exchanged genes with everything else to some extent, probably by way of restriction enzymes [chemicals that act like scissors in the cutting and splicing of genes] found in microorganisms. Including animals with plants, and plants with bacteria, and bacteria with animals. And this is happening all the time, though you can't prove it directly. In most cases you don't give the acquirer a selective advantage. In most cases it's a disadvantage, and in the tremendous battle that's going on in the soil, either it's going to die, or it's going to get rid of its [new] gene as fast as it can."

Keeler adds:

"Again, we must be careful not to generalize about microbes. Genes can surely move around the bacterial world, but normally they probably don't. Yes they can, but they probably don't. Rob Colwell's probably right: The average, innocuous event greatly outweighs the rare event or high-risk outlier. But that's really no comfort at all to the regulators."

Worst Case: Microbial Mayhem

So the worst-case scenario goes something like this: Put out in nature some mild-mannered bugs with some potent biological trick —

such as efficiently eating some toxic chemical — spliced into them and they compete pretty well. These gene-spliced bugs begin to produce methane as a by-product. Then they exchange genes in a Bacterial Bazaar, and the genes for toxic taste buds get passed to some tough local bacteria that now start punching like Mike Tyson. In turn, this starts a ripple effect of toxic chemical consumption and methane pollution that creates chaos in Bacterial Bazaars. Eventually, the big bacterial buffer that maintains all ecosystems begins to break down. Like an elegant, hand-crafted gown losing a thread, the whole thing frays and comes apart. And there is no replacement for Earth's microbial gown.

Whatever is said about possible microbial mayhem, the irony remains that no one will ever know ahead of time when or where a microbial mishap will occur. You can say this: It will be at the right place and the right time for the microbe, which doesn't give a hoot about human interpretations of its behavior; the right place and the wrong time for whoever deliberately released the microbe; and the wrong place at the right time for anyone opposed to releasing altered organisms, giving them fodder for their fight.

Compared to What?

Always ask: Why is a risk being taken? What is the benefit?

Just as "environment" became a center-stage global buzzword in the late 1980s, "risk" will soon become internationally famous as well. Because, to do anything about cleaning up the messes we have already made in the environment, something must be done. And whatever that something is, risk will be involved.

In the end, the best way to look at any risk — and the way all of us make decisions about risks every day — is why is the risk being taken? In short, what's the benefit? And what's the outcome if the risk is not taken?

Unless considered in this real-world context of alternatives, any risk can be made to look huge and unacceptable. Automobiles, a frightening thing for people who had never ridden in or driven one, became easier to accept when people realized they had gone farther, in less time and without saddle sores or sore feet. Considered by themselves, with maintenance, insurance, danger and so on, forget it; but up against the alternatives, well, when's the last time you rode on a horse, in a buggy or walked 20 miles to see a friend?

No one can deny that some organisms produced with biotechnologies introduce new risks. They do. Neither should anyone try to discount, much less explain, what the "Tally-the-Trauma Tote Board" reveals: Biotech, or genetic engineering, makes many people uneasy. It's unfamiliar, powerful, and being applied to organisms that are unfamiliar, potentially difficult — if not impossible — to control, and invisible.

But the most important question in the deliberate release debate is: What are the reasons businesses and researchers around the world are willing to tackle such formidable odds? Are those reasons valid? And what other risks are we taking by not learning more about microbial behavior outdoors?

An Ill We Can Bear?

One example of "compared to what?" will suffice. It comes from an exhaustive, revealing 1988 poll conducted among citizens of North Carolina by the highly regarded North Carolina Biotechnology Center. Having been told that scientists have developed different ways to fight plant diseases, respondents were asked how they felt about a farmer using:

- genetically engineered plants? 76% favored their use;
- genetically engineered bacteria? 54% favored their use;
- genetically engineered viruses? 44% favored their use;
- new chemicals? 28% favored their use.

Dr. Adrianne Massey of the Center's staff simply says: "A genetically engineered virus ahead of more chemicals? Let me tell you, that surprised us."

What's most surprising, really, is how seldom biotech's many potential applications outside have been compared to what they might replace and/or clean up — chemical solutions that have *caused* problems that people everywhere are fed up with.

Hamlet was right: "There is a natural tendency to bear those ills we have, rather than to fly to others that we know not of." That is, until those ills we have become intolerable.

6 Regulating a Revolution

"When I saw the full scale of the disaster in Prince William Sound in Alaska ... my first thought was: Where are the exotic new technologies, the products of genetic engineering, that can help us clean this up?"

These dramatic words were delivered by EPA Administrator William Reilly in his first speech about biotechnology on October 13, 1989, to a Washington, D.C. audience.

"Where are the environmental solutions we so desperately need?" Reilly wanted to know. Then this clarion call: "Let us hear more about ways that biotechnology can help solve environmental problems, from oil spills to Superfund sites. I want your ideas, your suggestions, about how EPA can encourage these innovations."

Though stirring, Reilly's remarks resounded with irony. In the first place, oil-munching microbes already exist. As we saw on page 28, one such "exotic" bug was the first "product of genetic engineering" to be awarded a patent on June 16, 1980.

But the richer irony in Reilly's remarks was this: Had he decided to use a biotech bug to help clean up Alaskan beaches, his own agency — the lead agency in the U.S. government when it comes to regulating releases of microbes — would have been a major obstacle in his way. *Ten years after that first microbe was patented, the EPA still has no clear rules regarding the large-scale release of genetically engineered microbes.*

What's more, had Reilly tried to use *biotech* "bioremediation" — the use of microbes to clean up the environment — he would have been swarmed by protest from leading environmental groups that defend and protect such pristine places as Prince William Sound. The National Wildlife Federation, Environmental Defense Fund, National Audubon Society and others are all ready to resist the release of *any* genetically engineered microbe on large scales until the federal government has a clear set of rules in place.

What's Going on Here?

Reilly did, however, order the experimental use of microbes to help in the Valdez disaster relief; in fact, he called this successful use of bioremediation "virtually the only good news to come out of that tragic situation." And not one environmental group raised a voice in protest. Why? Reilly: "The Alaskan spill cleanup has been a local-hire enterprise from the start, and *all our bacteria are native Alaskans.*" [emphasis added]

The crucial fact: Reilly did not dare employ *genetically engineered* microbes.

Regulatory Storms

Definition — White squall: A tropical whirlwind, coming on without warning other than a small white cloud.

The whole notion of deliberately releasing genetically engineered microbes hit the regulatory world in the early 1980s like a squall — but with scientists' white lab coats instead of white warning clouds. And ever since, the issues kicked up by deliberate-release squalls have been giving regulators major headaches.

This is hardly surprising, as Dr. Ralph Hardy, president of the Boyce Thompson Institute for Plant Research at Cornell University, explained in 1987:

"Biotechnology is moving fast, very fast; in fact, much faster than I originally estimated. I'm now convinced that we will generate more scientific information between now and the year 2000 than has been generated throughout all human history."

Regulators can't predict what squalls biotech will produce next and, when they involve releasing microbes, what impact they might have on the world's ecosystems — let alone get a regulatory net around volatile scientific and media storms to coax them down the path of least disruption and maximum benefit toward public confidence.

Welcome to the task that William Reilly now faces: regulating biotech squalls as they move outdoors. To understand the complexity of this task, we must briefly return to the mid-1970s.

The Big Engine that Could

"Such is the human condition: The trains start arriving when we are in the early stages of thinking about how to build a railway station."

from *To Govern Evolution* (1987), by Walter Truett Anderson

Remember when those 100 or so scientists met at Asilomar, Calif. — you know, the ones who not only raised a red flag over biotech, but also put the regulatory train on biotech's tracks. (See page 46.) Well, the engine they built was concern over "genetic engineering."

In 1976, some of those same scientists-turned-regulators drafted the world's first rules governing biotech — creating, in effect, the first car behind the engine of concern pulling the biotech regulatory train. Known as the National Institutes of Health (NIH) guidelines, those first rules applied only to federally funded research and thus had no teeth to bite anyone who was not dipping into Uncle Sam's wallet. Industry compliance was voluntary.

Dr. Elizabeth Milewski, who served for three years at NIH as assistant to the director of the Office of Recombinant DNA Activities and thus knows the guidelines' history as well as anyone, makes clear that they arose purely because of the *process* of biotech:

"Certainly there was process focus from Asilomar that carried into the early '80s... That biotech is the extension of a very old process is a standard piece we all put in our writings, but it was the advent of recombinant DNA that led to the NIH guidelines for research involving rDNA molecules."

Genetic engineering so stimulated public nightmares of mutant monsters inadvertently released from labs across the land that the very first regulatory message to the public was, in effect, "fear not — biotech can be successfully contained *indoors.*" Indeed, the impetus behind and exclusive focus of the NIH guidelines was the proper containment of the processes and products of rDNA research. John Cohrssen, an attorney for the President's Council on Environmental Quality, said in June 1988:

"The NIH guidelines assumed that additional risk was added to any organism created with biotech because of the biotech. Never had — nor has — any evidence supported this implicit assumption, but it created a negative perception that many of us in government are still trying to change."

Amazingly, almost overnight, the NIH guidelines became the foundation for international regulation of biotech. (See Chapter 7, beginning on page 135, for further discussion of how other countries are reacting to and regulating biotech.)

Release a No-No

Attorney Robert Nicholas, a former senior congressional official who is a veteran of U.S. regulatory wars on biotech since the 1970s and who now represents many biotechnology companies in Washington, D.C., points out the key facts for the deliberate release debate in the 1970s:

1. The NIH guidelines prohibited any release of rDNA organisms; but

2. They authorized the director of NIH to approve releases when the scientific evidence so warranted.

"You must remember," Nicholas explains, "in the mid-'70s, biotech was still mostly an academic pursuit focused mainly on human health and not yet driven as it would be in the 1980s by the need to get outside and test experiments that might become products. At that time, environmental introduction was not a major step in most research. So prohibiting deliberate releases of genetically modified organisms was seen as a politically smart step toward enhancing public confidence in biotechnology and its regulation. Something would have to change when researchers wanted to go outside to test their inventions in the environment."

Something did change.

U.C. Berkeley scientist Steven Lindow, whom we met on page 51, asked NIH to let him test ice minus bacteria outdoors. After five years of intense focus on keeping biotech properly contained inside and taking great lengths to prevent "it" from getting outside, suddenly someone was knocking at the lab door, wanting to put "it" outside.

Few people recognized at the time that the engine of concern that was pulling the biotech train also was pushing public perception of agricultural research and development in a new direction. Before biotech, Lindow would have been asking to "introduce" his altered organisms into the environment — for "introduction" describes the entire history of the development of agriculture (see page 8).* But after genetic engineering, Lindow was asking to "deliberately release" his microbes.

John Cohrssen points out the important shift in perception represented by this new nomenclature:

"We got saddled with the concept of deliberate release early and haven't been able to shake it. It's a misnomer that carries a very different connotation and feeling from introduction."

*For a fascinating history of plant introductions into the United States and around the world, see CSI's *BriefBook: Biotechnology and Genetic Diversity*. See inside front cover for details.

Indeed, as anyone who has ever babysat dogs, cats or kids knows, putting something outside intentionally, when you've been told for five years that under *no* circumstances should the beasts be allowed to get out, can cause anxieties to rise and tempers to flare.

Down came that avalanche of angst that buried Lindow for five years (see page 51) and up came the first of many biotech-generated regulatory squalls throughout the 1980s.

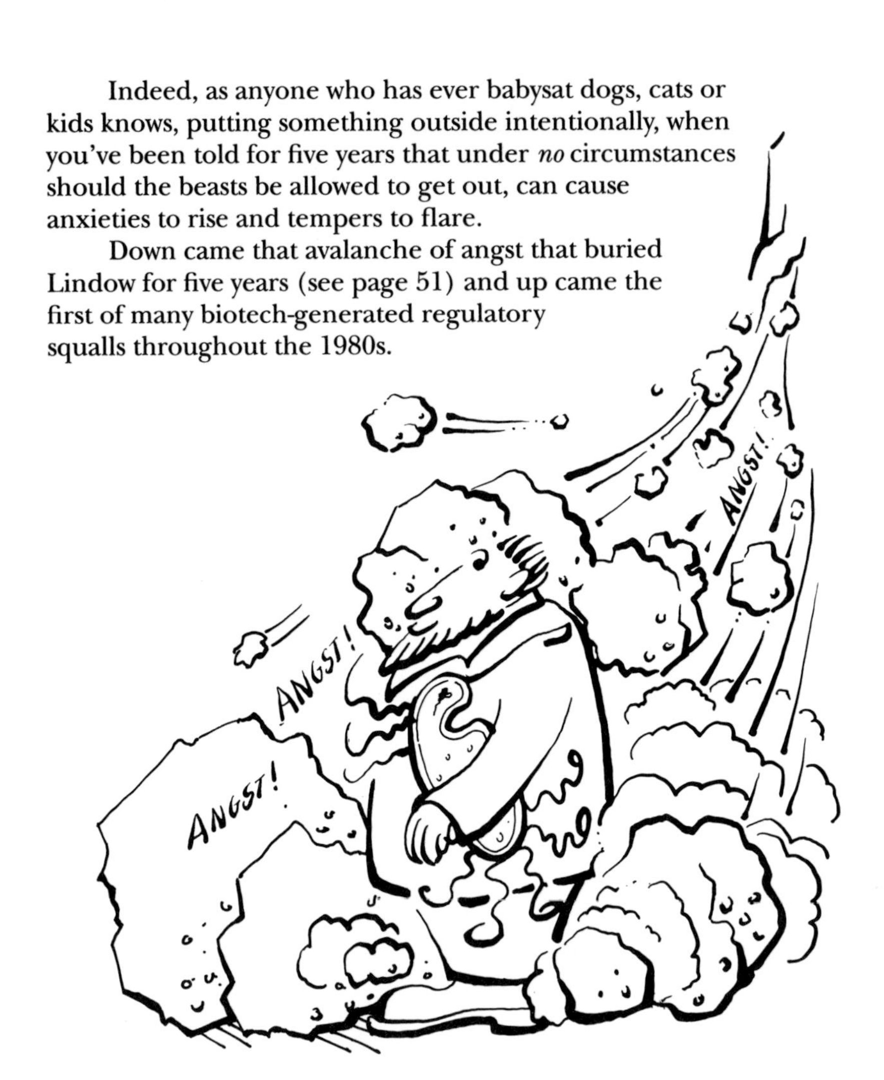

Caught without a Clue

The media squall caused by Lindow's request caught Washington's regulators completely off guard. Well into the early 1980s, the regulatory agencies hadn't a clue as to *whether* they were going to regulate biotech as it moved outdoors, much less how. And, as predictably as parents publicly embarrassed by their charges' behavior, Congress began verbally spanking regulatory agencies in media-hyped hearings.

Nicholas, former chief of staff for then-Rep. Albert Gore, D-Tenn., says he organized those congressional hearings with three purposes:

1. To ask: Who's minding the outdoor store? Are any federal regulators evaluating biotech as it moves into the environment?
2. To determine who should be regulating that part of the biotech revolution that has the potential to affect the environment; and
3. To facilitate the commercial development of biotech by providing a clear and simple regulatory roadmap for everyone, companies and academic institutions, to follow.

The early answer from agencies on who's going to regulate biotech was: everyone who can.

You see, it's not often that a huge and complex technological whale swims into Washington seemingly unnoticed by all. But as soon as someone hollers, "Thar she blows," everybody heads to the scene for a piece of the prize. So it was when Lindow's request reverberated across Washington — all those federal regulators who wanted a chunk of the biotech whale started to assess whether their laws formed a regulatory net that could land one.

It wasn't just regulators, but industry, academics, environmentalists, trade organizations, activists and the media...suddenly everyone wanted to regulate the products of biotech. Dr. David Kingsbury, who led the effort to slice up the biotech whale for President Ronald Reagan during much of the 1980s, looks back:

"The period from late 1983 to 1986 really was one of hysteria. Everyone was feeling the pressure from the Hill [Congress] to **do** *something, but a lot of us really didn't understand the ramifications of regulation. I could sense that a bureaucracy was brewing, but it was like a juggernaut — things were moving and taking on a life of their own. Congressional pressure and turf wars among agencies were growing acute and a lot of us did not understand the implications."*

Then one agency seized the moment.

EPA in the Driver's Seat

On October 17, 1984, the EPA announced in the *Federal Register* — the publication that Washington-watchers follow as raptly as stockbrokers scan their video screens — that it would "require notification prior to all small-scale field tests involving certain microbial pesticides in order to determine whether experimental use permits are required." Until that time, EPA had "presumed" under regulations of the Federal Insecticide, Fungicide and Rodenticide Act or FIFRA, which Congress enacted in 1947 to register pesticides, that pesticide manufacturers could test their products outdoors on less than 10 acres of land or 1 acre of water without having to notify EPA. That's because, EPA says, "Chemical pesticides have no independent mobility or reproductive capability."

But microbes are *alive* and have both. Thus EPA announced, in essence, that for microbial pesticides, *do not,* you biotech-bugmakers, presume that it's O.K. to test your microbes outside on small scales without notifying EPA. Any release of microbial pesticides, EPA's logic goes, is significant and EPA should be notified to determine whether a full-blown review is required.

With this seemingly small action, EPA instantly guaranteed itself most of the car called "microbial releases" in the biotech regulatory train.

Cohrssen claims, "EPA made a corporate decision that the area of microbes was ripe for regulation and they were going to make that their business." Dr. Henry Miller, who sits atop his own biotech empire as director of FDA's Office of Biotechnology, says bluntly:

"It was a desire to create an empire. It's like Monopoly. It's better to have three hotels than it is to have one hotel."

EPA's minor move was a big straw placed on the already overloaded back of a public primed to protest early biotech releases. It helped create a new phenomenon: small-scale field trials that were highly scrutinized media events. And these events literally defined the mid-1980s for microbial releases. Before October 1984, outside tests of experimental microbes — or chemicals, or plants, or animals or most anything else except nuclear devices — had never been news.

EPA helped change that, as David Kingsbury, now a professor at George Washington University Medical School, notes with clear regret:

"If EPA hadn't done that, if more of us had sensed the magnitude of a decision that, in the frenzy of activity at the time, literally slipped past everyone, the world would be different now. That was a ***very*** *big step."*

It was a turning point in the release of genetically engineered microbes. After all, at that time, most of the world's regulators were watching how U.S. agencies were groping with releases, so the EPA focused attention on the notion of "deliberate release." And it was this decision by EPA that would have made it nearly impossible for its current director to use, had they been available, genetically engineered microbes to clean up Exxon's colossal blunder in Alaska in March 1989.

When will the U.S. biotech train push past small-scale releases into the bumpy frontiers of large-scale releases? No one really knows yet.

But read on.

Seats on the Train

EPA's October surprise made other agencies eager to jump on board the biotech train. Most did when the first attempt at a "Coordinated Framework" for federal regulation of biotech was published in the *Federal Register* in December 1984. The Coordinated Framework was everyone's "first cut" at where they might sit once aboard. Eighteen federal agencies and executive offices had searched their statutes to find the correct phrasing to win themselves first-class seats. With so many agencies in the hunt, the result, according to Nicholas, was:

"Incomprehensible, overlapping and inconsistent jurisdictions. But that's true with the regulation of any new technology, especially one as sweeping as biotech. It encompasses so many potential applications — from food to mining to pollution clean-up to cosmetics to livestock to drugs and vaccines, to name a few — most of which are already regulated as end products by a patchwork of laws and regulations. And that means it stretches across many agencies."

Jack Doyle, director of Friends of the Earth's Agriculture and Biotechnology Project, speaks for many critics of the first framework when he says:

"The original framework was trashed by most everybody. EPA was and is the logical choice as the lead agency, because it is independent and its specific charge is to protect the environment. But turf wars had already begun and no one lets go easily."

So it was back to the drawing board — countless more "inter-agency" meetings and memos — to improve the Coordinated Framework.

Framework Take Two

Eighteen months passed before the executive branch published on June 26, 1986, the "final part of Coordinated Framework for the Regulation of Biotechnology" in the *Federal Register.* Though much had changed from the 1984 version, the basic configuration of the biotech train remained the same. If your biotech product is ...

- ... going outside and you can't see it with the naked eye — namely, microbes of any kind — you must start in the EPA car.
- ... going outside and you can see or hear it — namely, plants and some animals — you must start in the USDA Animal and Plant Health Inspection Service (APHIS) car.
- ... going into humans or animals — namely, food and medical products — you must start in the FDA car.
- ... going to continue being funded by the federal government — namely, any biotech research that's federally funded — you're in the NIH car, and possibly in the USDA car too.

These basic divisions stand to this day, but obviously overlaps occur. A good example: while animal drugs are covered by the FDA, animal biologics* fall under USDA/APHIS oversight. Why aren't these

*Biologics are any biological products, such as vaccines, used to induce immunity to infectious diseases.

two regulated by one agency as common sense suggests? Nicholas nails the reason:

"Various congressional committees oversee different agencies and thus different statutes and the products they control. They protect their turf just as agencies protect theirs. Congressional committees drive lots of these jurisdictional turf wars and illogical interpretations of statutory reach. The only way real federal coordination of biotech can occur is when some committees become willing to give up pieces of power — and that just doesn't happen in Washington."

Statutory Stretch Marks

All of the statutes currently used to regulate biotech were written before biotech even existed.

Under the final framework, while EPA controlled most of the car called microbial releases, USDA/APHIS had some seats reserved as well.

With the "who" question answered, the next question was how?

The trick to understanding how anything gets regulated in Washington is to understand the laws that apply — or don't directly apply but with a little stretching or "interpretation" or even some actual changes or "rulemaking" can be made to apply. When you realize that all of the statutes currently used to regulate biotech were written *before biotech even existed,* you can see right away that, by definition, biotech regulation has involved plenty of "statutory interpretation." In fact, biotech has given some statutes stretch marks.

Elizabeth Milewski, who, as special assistant for biotechnology at EPA, has been central to EPA's effort to regulate genetically engineered microbes since it began, explains the "trick" to statute stretching:

"The trick with an existing statute is to use its powers and its defining language and translate it into the world of microorganisms, the world of biology."

A closer look at the laws the two leading players use to regulate microbial releases will reveal that, even though some statutes have been stretched to their limits like bungie cords wrapped around a whale, most observers believe they are doing the job. Even so, some people would just as soon see them snap, forcing the federal government to start over and draft new laws that better fit the monumental task of regulating something so big.

Two Big Bungie Cords

EPA has two major statutes it stretches around biotech.

The first, FIFRA, we saw on page 116 and it requires the least stretching for biotech products. It has been used for decades to register microbial pesticides based primarily on naturally occurring Bt strains. The definition EPA uses for pesticide is quite broad and, if "interpreted" just so, might even include a stiff cocktail consumed to "mitigate" the effects of one's in-laws. It reads:

"Any substance or mixture of substances intended for preventing, destroying, repelling or mitigating any pest and any substance or mixture of substances intended for use as a plant regulator, defoliant, or desiccant."

FIFRA originally was administered by the USDA, but was transferred to EPA upon EPA's creation in 1970 because, as House Agriculture Committee Consultant Skip Stiles says, "the widely held opinion at the time was that USDA had hardly developed a reputation as a rigid regulator of anything, much less agricultural pesticides."

EPA's second major "biotech" statute, the Toxic Substances Control Act or TSCA (pronounced "TOSCA"), took effect on January 1,1977, as EPA's major gap-filler in chemical regulation. It covers all "chemical substances" in commerce or intended for commerce not already under other statutes — including genetically engineered microbes. As Milewski explains:

"In the early '80s EPA made an interpretation that since living organisms are at basis chemical substances, those living organisms could be subject to TSCA."

Though TSCA — like FIFRA — was written exclusively with *non-living chemicals* in mind, calling microbes chemicals is correct on one level: Genes are made from chemical building blocks.

But on another level, calling microbes chemicals is tantamount to calling chemists chemicals — or regulators chemicals. Milewski, a molecular biologist herself, is the first to admit the odd and imperfect nature of this application of TSCA:

"Microbes are clearly not the same as chemicals. They are alive and reproduce. They change depending on the environment you place them in, and

they change even if you don't move them at all. Chemicals clearly aren't like that."

Words without Definitions

If defining microbes as chemicals was an awkward way to establish TSCA's role in regulating any non-pesticidal — non-FIFRA — microbes, then the terms EPA chose to delineate which microbes TSCA would "capture" for review were even clumsier. In the 1986 framework, EPA said it would review any "intergeneric" microbe — one that contains genes from organisms in different genera — and any microbe that contains materials derived from "pathogenic" organisms — those known to cause harm to other organisms.

The first term, "intergeneric," seemed logical enough. After all, biotech is the only force capable in most instances of combining genes from different genera. Such organisms are usually unique, and "new" is a key trigger in TSCA's defining language. Milewski elaborates:

"It's a hell of a lot easier to decide what's a new chemical. You add a new chemical group onto a compound and you've got a new chemical. With a microorganism it's more difficult. We had to choose something, so we chose intergeneric. We said if you combine the DNA from organisms that are classified in different taxonomic genera, you are subject to TSCA. Because it's a 'new' microorganism."

The second term, "pathogenic," proved downright impossible to define. In fact, EPA never has come up with a definition of pathogen that works for its purposes or anyone else's for that matter. That's partly because an organism's pathogenicity can depend upon where it winds up. *E. coli,* for example, is part of every human's intestinal flora; but *E. coli* also causes certain eye and urinary tract infections. What's more, referring to EPA's problems with defining "pathogenic," Terry Medley, director of the Biotechnology, Biologics and Environmental Protection Unit at APHIS, points out:

"It's very hard to define pathogen because normally when people think about pathogen they think about something that causes disease. But it's more than just disease; it's injury, damage and so on."

In its 1986 policy statement, EPA tried to make its regulations for genetically engineered microbes under TSCA product-based because industry, regulators and scientists were trying to get away from the notion of process-based regulation. But, as one top-level EPA veteran admits:

"We all say it's product-based just to keep everyone off our backs, but often use process because you just can't draw the line. Biotech, the process, is the whole reason all the biotech conferences, the biotech critics and the biotech businesses exist. It's what the public is worried about and what we have to give the appearance we're regulating."

The World's a Pest

Some gumball philosopher once said that if you're a hammer, then all the world looks like a nail. Well, if you're working for USDA/APHIS, then all the world looks like a plant pest.

You see, the statute APHIS uses to claim its share of the seats in the microbes car on the biotech train is called the Federal Plant Pest Act (PPA). Passed by Congress on May 23, 1957, PPA enabled APHIS to claim in 1984 that, as Medley makes plain: "If a microbe presents any potential for plant pest risk, then it will be regulated by APHIS."

How broad a net can the words "plant pest" cast? "Extremely broad," Medley says. And so suggests the actual definition from the 1957 Act:

"'Plant pest' means any living stage of: Any insects, mites, nematodes, slugs, snails, protozoa, or other invertebrate animals, bacteria, fungi, other parasitic plants or reproductive parts thereof, viruses, or any organisms similar to or allied with any of the foregoing, or any infectious substances, which can directly or indirectly injure or cause disease or damage in any plants or parts thereof, or any processed, manufactured, or other products of plants."

Medley, who wrote the adaptation to APHIS rules for genetically engineered organisms that was published in the *Federal Register* in June 1987, goes on:

"'Plant pest' potentially includes synthetic DNA and unknown and unclassified materials. On top of that, another clause in the Act says that even if something is not on our list and we have reason to believe it might cause plant pest harm, then we can take action on it."

Summing up, Medley puts his "you-name-it-and-it's-a-potential-plant-pest" act in perspective:

"Congress intended the act to be 'gap-filling authority' that enabled USDA to become the first line of defense to protect American agriculture from plant pests. It's like TSCA in this sense, because TSCA is 'gap-filling authority' for EPA to catch and deal with any toxic chemicals that other statutes miss."

One big problem with stretching statutes so far is that things that clearly are not plant pests are reviewed as "potential plant pests" because that's the way the PPA regulates. As Calgene Director of Regulatory Affairs Don Emlay says, "It's unnecessary name-calling, but we understand why it's done."

It's also the trick to statute stretching. No one ever said it was pretty — or easy to understand.

Loophole: Microbes for Money Only

Definitions, it turns out, are just part of EPA's problems. Another is scope: how big a chunk of the biotech whale its two main statutes can be stretched around. Particularly with TSCA, this is not at all clear. TSCA is a gap-filling statute that EPA can use to catch all those microbes not nabbed by FIFRA or by APHIS's Plant Pest Act. It is the federal net for microbes under all the others, designed to catch anything that falls through. But it has some "big, obvious holes," says NWF's Margaret Mellon.

The biggest hole is that TSCA only applies to commercial research and therefore currently does not cover microbes used in academic research. Whether they're designed for pollution clean-up, mineral extraction, waste treatment, biomass degradation, fertilization ... if microbes aren't made for producing money they fall through the TSCA net. If, say, some genetically engineered microbes can munch oil spills and only the environment stands to gain from their use, they would fall through the TSCA net for review. Yet no one would dare release a biotech bug without federal review — especially the federal government.

Nevertheless, EPA's Evans, in a speech explaining TSCA to state regulators in June 1989, said:

"We at the agency feel that biotechnology is going to be very beneficial; we see it being used aggressively in waste treatment areas particularly."

Notwithstanding Evan's statement, biotech won't be used to clean up anything outside until TSCA gets fixed. And that won't be anytime soon if Mellon and other critics of the federal framework have their way. Mellon feels strongly that TSCA should be scrapped for use on microbes:

"TSCA doesn't really work at all for chemicals, so using it to regulate microbes as chemicals make no sense. It's time that the U.S. step back and take a hard look at the framework. When we do, I think the public will demand a regulatory system designed specifically for the unique, living products of biotech. It's the only thing that makes sense."

Sticky Issues and Politics

EPA's been trying to patch up TSCA since 1986, but it hasn't gotten far. In fact, because of bureaucratic delays caused by so many people trying to pull TSCA in different directions — if not simply pull it apart — EPA is unlikely to publish a "final rule" revamping TSCA before 1992.

Mellon calls this near-decade delay "just outrageous."

Why is it taking so long? Why doesn't the world's leading country in all forms of genetically engineered microbes destined for use outdoors have a clear regulatory system in place for these microbes? First, because of some plain ol' sticky issues that no one can agree on. And second, because of what defines Washington — politics.

The top four sticky issues blocking a final TSCA rule are:

1. How do you distinguish between commercial research and development (R&D) and academic R&D?
2. Should TSCA cover microbial releases for *any* research and development purposes?

3. Should TSCA cover significant new uses of *non*-genetically engineered microbes on large scales?

4. When does small-scale release become large-scale?

Let's take them in order.

1. The first issue is a definitional dilemma. Where does academic research end and commercial research begin? Currently, nobody can answer that question. But, as Sandoz Corp.'s director of plant biotechnology research, Dr. Ron Meeusen, says: "Biotech has blurred, if not blown away, most lines between the two. Things were a lot simpler to define and to distinguish before biotech."

2. The second issue causes academics to cringe because most academic researchers hate spending precious time and grant monies trying to figure out regulations, especially ones as complex as the Coordinated Framework. Elizabeth Milewski believes:

"Academics are the real wild card in all this. Industry and public interest groups predictably disagree, but they're at least clear on the need for regulations and how the process works. Most academics have no experience with regulations and that's just fine with them."

3. The third sticky issue takes us back to the 1988 Winter Olympics in Calgary, Canada. Remember that silly scenario on page 43 about Snomax freezing all of Canada? Well, Eastman Kodak's Snomax Technologies used Snowmax to seed clouds in 1989. You ask, "Isn't that a 'significant new use?'" Not to EPA. The dead microbes in Snomax are treated as chemicals and thus exempted for small-scale tests. EPA has already said Snomax is safe and we know it's not genetically engineered, so, whoosh, nothing hooks it.

Indeed, companies shudder at the prospect of regulations for natural microbes, as do organic farmers and anyone interested in bioremediation. The use of these non-engineered microbes, made by nature, would virtually dry up if the often prohibitive costs of regulatory approval were added to their cost of development.

4. Finally, what is large scale? So far, because of EPA's October 1984 surprise, it's *de facto* more than 10 acres of land or one acre of water.

But as yet no one has actually *defined* what constitutes large scale. Someone will have to, and soon, as companies like CGI continue to force this issue to the fore (see page 82). Mycogen Chairman Jerry Caulder complains:

"I've had a devil of a time getting EPA to approve a 710-acre field trial with a biopesticide that's delivered in 'dead' microbes — even though we have been testing, with EPA approval, the same system on small scales for four years."

You're getting a feel for why regulating biotech is giving many people in Washington, D.C. and other capitals around the world major headaches.

Ent 'OMB'ed

EPA officials had a refurbished version of TSCA ready for public review in May 1988, but then they sent it to the Office of Management and Budget (OMB). All proposed regulatory changes must be reviewed by OMB, primarily for potential budgetary impacts.

Enter politics. Exit TSCA version #2.

OMB sat on EPA's rule for nine months and it died. In a classic behind-the-scenes Washington battle — involving EPA, the White House, OMB, FDA, NIH, industry and other agencies — venom flowed freely for months. Reagan administration EPA Administrator Lee Thomas, Milewski claims, "was very eager to try to get some type of resolution of biotech rules before he left, before the change of administration." But powerful people in the Reagan administration decided otherwise.

What counts in Washington is less who the players were and more what remains. And what remains is the 1986 framework that Robert Nicholas suggests "has had a much longer shelf life than anyone imagined or probably intended." Maureen Hinkle says the framework was "crafted by some very creative bureaucrats — they had to be, because they didn't have very much to work with." And Margaret Mellon just shakes her head and calls the framework as it stands "a wobbly and incomplete edifice."

Will it Stand?...

Can the federal framework withstand the dual weight of rapidly developing technology and increasing public scrutiny?

Most observers think it will. Orlo Ehart, who has worked with the framework both as a regulator for the state of Wisconsin and as a regulatory specialist for CIBA-GEIGY Corp., says frankly, "It will never be clear to anyone except biotech watchers and we're a small crowd." In other words, decision-makers in Washington — and elsewhere — must first understand the federal framework before they can recognize either its weaknesses or its strengths. Only then can they consider remodeling it — much less scrapping it and starting over.

And understanding the framework is no easy task. U.S. Senate-staff biotech watcher Kathleen Merrigan candidly admits what she and others in the policy-making business are up against:

"I was trying to brief three interested senators on microbes and, when I started talking about horizontal gene transfers, their eyes just glazed over. One just outright laughed and said, 'What are you talking about?'

"The real question is, how do you get important people involved when the material is just not accessible? Microbes just don't hold up against other issues like the deficit, drug wars and freeway systems that need repair. Biotech is one of those issues that overwhelms the political machinery here in Washington. And the framework doesn't help. After a few years I still don't fully understand it; it's a very complex structure."

One result of both the framework's and biotech's complexity is that, as California Rep. George Brown, one of the best-informed leaders in Washington on biotech, described it in a 1987 speech on biotech: "It is difficult to discern a prevailing view in Congress on this issue."

No wonder. Congress waits to address most issues until they flare like brushfires; *preventing* brushfires is not Congress's forte. Smokey the Bear may be alive and well in the backwoods, but not on the front lines of legislation. And since nothing really has gone wrong with biotech that would force Congress to focus on it, it stays "down around priority No. 57 on most leaders' lists," according to Dr. Ron Cape, founder and chairman of Cetus Corp. in Emeryville, Calif.

Good news remains no news.

... Odds are it Will

However complex, confusing and "wobbly" the Coordinated Framework may be, to date Congress has shown little interest in remodeling or replacing it. This is because none of the proposed alternatives has looked any better. Indeed, Congress has yet to pass any biotech bill regarding releases of genetically engineered microbes — or any other major part of the framework.

This fact frustrates activist Jack Doyle, who asserts:

"This technology is so important for the natural resource industries and therefore for the environment, surely the proper forum for policy discussion is not Bill Reilly's office — it's the Congress."

Nicholas observes:

"While cumbersome, the framework remains flexible and has worked reasonably well so far. As the NIH guidelines were, the framework will likely be relaxed in the coming years. New 'biotech' legislation and/or a new 'biotech agency' would doubtless create problems of their own, so I just don't see either happening. After all, if an easy solution were available, Congress would find the will and a vehicle to get the job done."

What are the chances that Congress will become interested in biotech regulation in a big way? Most observers say not very good. Framework foe Mellon says with regret:

"Congress always puts off preventative measures in favor of crises focus. They say, 'Not in my tenure; let the next group deal with it.' It's lousy statesmanship, but it happens all the time."

So, unless a major biotech brushfire breaks out — unless something goes wrong, the framework will stand. However, as Rep. Brown said in 1987:

"If the slightest thing indicates that the Coordinated Framework isn't working, Congress will jump in there and create a new one."

Some States Won't Wait

Newsweek magazine, that siren of America's mood, saw it in November 1989 (11/13/89):

"A new age of environmental federalism has dawned. In a stunning switch, the states are no longer merely implementing federal standards but are setting the environmental agenda."

Skip Stiles, who has worked on The Hill for 15 years, saw it coming in mid-1988:

"The good news is that the Reagan revolution worked: Federal regulation at all levels has been rolled back. The bad news is that the Reagan revolution worked: There's a growing sense that the feds are out of control and have lost touch with Americans' desire to be safe and clean. Scandals at HUD, FDA, and the savings and loans; EPA's inability to deal with pesticides — across the board you're seeing signs of Reagan's cutbacks.

"All this comes at precisely the time that biotech firms need to get their products to market and prove that biotech's for real. But the irony is that industry can't move out the door with any speed or certainty, at least not with microbes. Many in the biotech industry have lost sight of the fact that a strong federal regulatory system ***lowers*** *costs in the long term. The price of perceived safety is invaluable.*

"And, as Reagan promised, regulation is moving to the state level and the public starts to lose the feeling that the federal government is adequately protecting them."

Does this trend toward environmental regulation moving to the states affect biotech? Most certainly. In fact, more biotech bills were introduced in state legislatures in 1989 than in all previous years combined. Does this mean that states will soon take over biotech regulation and leave the federal framework a hollow shell? Hardly.

R. Steven Brown, who tracks state biotech legislative activity for the Council on State Governments, says: "State legislators see biotech as an economic development issue first and an environmental issue second." But the environment leapfrogs into first when a release is proposed, public awareness is whipped up and state regulators are

Most state legislators see biotech as an economic development issue first.

forced to do a crash course in the federal framework. At first many of them find it maddening, as Ehart recalls:

"When I first saw the federal framework from the vantage point of a state regulator, I thought it stunk. But then I realized that Congress has done nothing to articulate policy. So the federal agencies are forced to live within their existing statutory boundaries and do the best they can with the new processes and products biotech produces."

That's not good enough for states that recognize the real problem is not that federal regulators can't regulate biotech's *products*. Indeed, as well as anyone, they can. They simply don't have the mechanisms to deal with public perception of — preoccupation with, really — the *process* of genetic engineering. Ehart concludes:

"So federal agencies have the authority to look at products, but the public is concerned with the process. Agencies have to look at what the public is concerned about — the process. If they don't, then the states will."

And those states in which deliberate releases have been conducted or proposed are looking at the process. After all, when any release is proposed, those directly affected by it, the local folks, turn to those with whom they have direct contact — state and local representatives and agencies — to find answers.

Distrust Brings Biotech Closer to the People

"All politics is local."

Thomas "Tip" O'Neill, former speaker of the U.S. House of Representatives

Since 1983, nine states — North Carolina, Minnesota, Wisconsin, Hawaii, Michigan, New York, Oregon, Rhode Island and California — have passed legislation that pertains to deliberate releases of genetically engineered organisms. Minnesota and Wisconsin passed biotech legislation because of debates over specific proposed releases. In Minnesota's case, the move to add another layer of regulatory oversight to the federal framework was also motivated, says Sheldon Mains of Minnesota's State Planning Agency, "because of profound mistrust by many of our legislators of the federal process."

The Environmental Defense Fund's state director in North Carolina, Steve Levitas, has a similar view of what happened in his state:

"Regardless of the federal framework, North Carolina would still have acted. The federal system is limited, too tough to understand and allows only limited access — it's huge and slow-moving. And it's more subject to ideology and politics. States of course are not immune to either, but it's worse at the federal level. Also, there's profound mistrust of Washington in our region. For whatever reason, it's true."

All Eyes on North Carolina

North Carolina is the state that has received the most attention for its legislation on deliberate release. There are two explanations for this attention.

First, North Carolina's Genetically Engineered Organisms Act, which passed in August 1989, had the support of industry — or at least some of industry. The two trade groups representing the biotech industry — the Industrial Biotechnology Association (IBA) and the Association of Biotechnology Companies (ABC) — took opposite sides on the bill. IBA opposed it, ABC supported it. IBA President Dick Godown made clear his organization's position:

"As an industry, we are striving to prevent a patchwork of laws and their inevitable maze of bureaucratic paper work that could slow down innovation."

At the same time, one of IBA's premier members, CIBA-GEIGY Corp., was an important backer of North Carolina's bill because its agricultural biotech center is based in the state. What can be made of this split in industry? Well, industry wants stable, predictable regulations that it can factor into its cost of doing business. So, sensing the momentum building in North Carolina for a state voice in release decisions, those businesses with the most to lose from restrictive regulations in North Carolina got involved, hoping to come out with the best deal possible.

Debates Move to States

The second and most important reason that all eyes are on North Carolina's Genetically Engineered Organisms Act is that environmental groups are eager to see other states follow suit. EDF's Goldburg says:

"The North Carolina Act is the first model bill. It's a good example of balanced regulation that we hope other states will pick up and use."

Has North Carolina started a trend? Mellon says, "If a few more states get active, it will have." Down that path lies industry's worst nightmare: 50 states each with its own version of biotech regulations. But long before that could ever happen, a big biotech brushfire would break out in Washington, D.C., quickly forcing biotech out of the "57th" spot on Congress's agenda and up the list.

Who would start that brushfire? Probably the biotech industry's powerful lobbyists, who have done such a "good job" convincing Congress that alternatives to the "ugly" Coordinated Framework are even "uglier" that they have inadvertently pushed the debate out of Washington into state legislatures. There they have far more to lose and far less influence. Skip Stiles recognizes the irony in this:

"Wouldn't it be amazing to see industry clamoring to the White House not demanding less regulation, but more regulatory certainty at the federal level. Then we will have come full circle from the Reagan revolution. That's the way Washington works — or doesn't work: in circles."

The Perennial Problem: Elusive Solutions

"Laws are like cobwebs. They catch small flies, but let wasps and hornets break through."

Jonathan Swift

When it comes to biotech, virtually every scientist you ask will tell you that it's almost impossible to create predictive laws that will actually prevent problems — especially with ubiquitous, mysterious and invisible microbes. Molecular biologist Milewski says frankly:

"I'm not sure that there is any way that we could catch a real disaster. Some things may be fairly obvious, like bringing a microorganism into the United States that is a problem in another country, citrus canker let's say. Those you may be able to stop. But I think there will still be some that will slip through the net because we just don't know enough about microorganisms and the way they interact and the way they behave, to be able to predict what's going to happen."

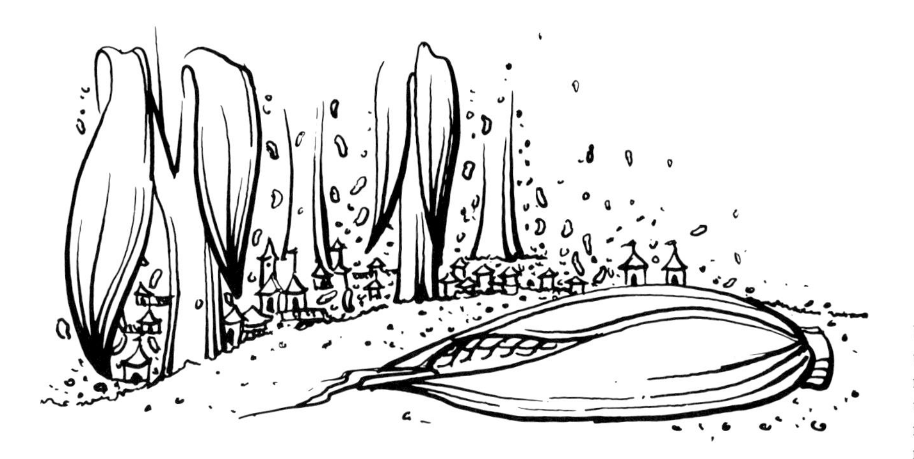

Understanding the mysteries of micospaces is critical to determining the real risks of releasing modified microbes.

What's a responsible regulator to do? Pretty much what they're already doing: reviewing every release on a "case-by-case basis" and requiring an extensive data trail so everybody can learn what data matter most when contemplating releases. Kathleen Merrigan says, "There are certain things that we can all agree on and one is that we need more information to have effective risk assessment on microbial releases."

Milewski concurs:

"I would much rather have some type of a data trail, even if it means extra time and money, than to have insufficient data trails or none at all. Even if we're wrong and a problem arises, at least we will have some knowledge of what was going on. Then your data base gets larger and your ability to understand what's going on with microorganisms increases."

Recalling his days on the front lines of release debates in Wisconsin, Orlo Ehart adds:

"Even the most ardent foes of field releases in Wisconsin eventually supported the process of field releases, in order to gain the data necessary to determine the real risks of release. It's the only way."

Perception is the Point ...

In any democracy, the public drives the legislators who pass the laws and set the policies that regulators enforce. So public perception is the key to any regulatory system. If the public perceives that a given regulatory system is working, then it won't demand change. Only when the public perceives that the store isn't being watched — usually when something goes wrong and makes the evening news — do regulatory systems get instant attention.

A good example: Few knew and fewer cared that supertankers were regulated, until the Exxon Valdez struck that reef in Prince William Sound. Suddenly, millions knew and cared. The store wasn't being watched.

Most people still don't know and don't care that microbes are regulated. So, as far as the public is concerned, the outdoor biotech store is being watched. Only flaring debates or unpredictable accidents will cause this perception to change. In outdoor biotech, as we have seen, a major cause of those debates is the principal subject of this *BriefBook:* the release of genetically engineered microbes into the environment.

... All Over the World

The same is true throughout most of the world. The public doesn't know much about biotech or microbes, much less how they are regulated. But they soon will, as everywhere people demand to know a lot more about their environment and what's being put into it.

Indeed, biotech will have to prove itself a part of, not separate from, the movement toward peace with the environment. If it doesn't, the use of genetically engineered microbes outside will remain just the stuff dreams, not products, are made of.

A Greening World Debates Releases

"If you don't have basically the same regulations governing the release of microbes in all countries over the face of the Earth, there's no point having any regulations at all. The Earth is too small. You can't autoclave people that come into Kennedy airport."*
French microbiologist Dr. David Tepfer

"If you try to control science and scientists with a book of rules, they will either go underground and do their work or simply go to another country that wants their expertise."
Dutch soil microbiologist Dr. J.G. Kuenen

"Govern the people by regulations...and they will flee from you...Govern them by moral force...and they will keep their self-respect and come to you of their own accord."
Confucius, Analects, II.3

Which two countries have superpower economies and technological wizardry that, in many cases, equals and even surpasses the United States? You know the answer: Japan and West Germany.

Now, guess which two countries are stalled on whether to deliberately release genetically engineered microbes? Japan and West Germany. That's right.

*An autoclave is a laboratory oven, of sorts, that is used to sterilize — kill all the microbes on — equipment, clothes, anything really, except humans.

Try this one. Which country has had by far the most deliberate releases of genetically engineered organisms per capita? Answer: Belgium.

How about this: Which Eastern-bloc country is welcoming Western companies that want to test their biotech products outdoors — as long as they bring their biotech expertise with them? The Soviet Union.

Finally, which country was the first to permit — with little review, minimal protest and almost no public awareness — a genetically engineered living microbe to be sold for use outdoors? Give up? Australia.

Around the World in a Chapter

Just as microbes are very small, the world is very big. And as you skip from country to country across the globe, it's tough to predict how each would react to the release of genetically engineered microbes.

This chapter will take you on a brief tour of countries grappling with the regulatory and media squalls generated by releases — or possible releases. It will introduce you to attitudes about releases that span a full and surprising spectrum.

Equally important is what this chapter won't do. It will not tell you precisely what releases have occurred where, nor will it give you a comprehensive look at the many regulatory approaches to releases that exist internationally. That's the stuff of not-so-brief books.

Most important, this chapter will not offer any sweeping conclusions about world attitudes on releases, because in this area — as in most — it's impossible to make global generalizations.

Some Rules of the Road

The following truisms about reactions to releases around the world will serve as rules of the road for our tour:

- Biotech has already gone global. It, like microbes, is everywhere. And anyone who applies biotechnology can easily create unique organisms that could be released, intentionally or otherwise, into environments whose occupants may not recognize the newcomers.

- Just because anyone or any country can create genetically engineered organisms doesn't mean that they will — and definitely doesn't mean that they will deliberately release those organisms. Japan is the best example: It has top-notch biotech capabilities, but a clear reluctance to date to release altered organisms, especially microbes, into the environment.
- Whether or not a country has an active environmental movement does not necessarily mean you can predict that country's attitude on biotech or deliberate releases. Denmark is very "green," meaning that it tends to be liberal on environmental issues, and has generally not supported deliberate releases. But Australia has a very active environmental movement and a prime minister known for aggressive environmental actions and policies, yet it has experienced little opposition to biotech or releases of altered organisms. And Belgium, definitely as green as countries get, is actively engaged in the release business.
- Only a fraction of nations worldwide — about 15 out of 158 — have actually released biotech-generated organisms — plants, animals or microbes.
- Finally, no country is currently pursuing, as part of a larger national strategy, a policy that includes releases for whatever purpose. Therefore, countries have, by definition, reacted to rather than planned for the issues that arise because of releases.

So what you wind up observing is less any country's actual policy on deliberate releases and more their reactions to releases, actual or contemplated. Therefore, the best a brief world tour can do is illustrate some of the major issues that accompany any release — and point out some of the cultural and political factors affecting releases, which vary enormously from country to country.

First stop on our whirlwind tour — Japan.

Japan, Where Anything Goes Indoors ...

Almost all of Japan's 250 firms involved with biotech became so after 1981. Ever since, biotech's corporate leaders in the United States have been sounding the same alarm heard from leaders of many other U.S.-born high-tech industries. The warning: We are in a race with the Japanese and we look a lot more like the hare in that old fable that's a turtle favorite. Cetus Corp. founder and chairman, Dr. Ron Cape, has said repeatedly since the early 1980s:

"The Japanese know they're in a crucially important race to develop biotech. They set national goals and focus on achieving them with tremendous intensity. What do we do? We act as if there's no way they'll pass us. Our leaders act as if they couldn't care less about biotech, as if there is no reason to worry. Let me tell ya folks, there is."

Monsanto President and CEO Earle Harbison put it this way to *Ag Biotechnology News* (10/89):

"In ag biotech, the Japanese are coming on strong... They understand something new and good when they see it. I don't think they go to bed."

Mycogen Chairman Jerry Caulder makes plain what the Japanese want from U.S. biotech companies: "Our technology, pure and simple. We develop it and they run with it." But, "If you sell your technology to Japan," as Akihiro Yoshikawa of the U.C. Berkeley Roundtable on the International Economy told *The Scientist* in September 1989, "chances are that Japan will come back and eat you up."

... Yet Nothing Goes Outdoors

Surprisingly, the Japanese are extremely reluctant to release genetically engineered microbes for any purpose into the environment. And as we have seen all along with microbes, like children, until you release 'em, you can't say how they will do in the big, bad world out there.

Why is a society that is so enamored with things technological and new so hesitant to try new microbes outside? *Shinjinrui.*

Literally, *shinjinrui* means "new human beings." In Japan, it means a new generation that isn't as enamored with things technological and new. Indeed, they're bringing with them as they move into power three things heretofore hard to find in Japan: a desire for leisure time, a yen for personal fulfillment and a budding concern for the global environment.

A good indication of this changing emphasis came in May 1989 when Japan's Environmental Agency coined the word "ecopolis" in its annual white paper on the environment. For years, the Japanese had boasted about building "technopolis," "science cities of the future." Then, suddenly, a prestigious government agency switches to seeking "ecopolis" — cities "where man lives in harmony with the environment."

As for attitudes toward biotech and specifically releases into the environment, revealing information came from a survey of readers of *Newton,* a very popular science magazine in Japan. Douglas McCormick, editor of *BioTechnology* magazine, said of the *Newton* survey:

"The results reflect not the general tenor of Japanese popular opinion, but the sentiments of a well-informed minority...which makes the results even more troubling."

The "troubling" results:

- 33% were uneasy about the prospect of biotech wines;
- 40% didn't trust vegetables bred in tissue culture;
- 72% were nervous about genetically engineered fish coming to market;
- 78% were concerned — with more than a third "very apprehensive"— to learn that genetically engineered microbes might be released in the United States;
- 90% rejected researchers' claims that releases can be environmentally safe — with a third willing to ban deliberate releases altogether.

These responses from Japan's science lovers make plain that no one should expect a wave of deliberate releases of altered microbes in Japan anytime soon.

To be sure, no one can say what the Japanese will do long term. When you consider that the Japanese government has targeted the clean-up of environmental pollution or "bioremediation" as part of its national program on biotech, you can't assume that they will remain reluctant to release microbes. And, as Cetus Corp.'s Cape cautions, "Don't count them out of anything that involves technology — ever."

West Germany Says, *Halt!*

West Germany is Europe's unrivaled economic leader. West Germany developed Europe's modern chemical industry. And West Germany is the country most opposed to the release of genetically engineered organisms. What's more, West Germany is the country most opposed to the *process* of genetic engineering for *any* purpose.

In a stunning 1989 decision, the Administrative Supreme Court of the state of Hesse blocked the multinational chemical company Hoechst AG from completing a plant in Frankfurt for the production of human insulin. Why? Because the plant technicians would be using *genetic engineering.*

They're baaaack — those two nerve-jangling words of the 1980s will remain controversial for the '90s in many parts of the world.

The Hesse court said that West German laws were insufficient when it comes to judging the permissibility of biotech "because of the different dimensions and the quality of risks connected to gene technology." Only the highest legal authority in the land, the Constitutional Supreme Court, can overrule the lower court's ruling, which remains binding on all states in the Federal Republic of Germany until and unless it is overturned. Or until the German government takes legislative action.

The courts and everyone else are waiting for the Bundestag, West Germany's parliament, to pass the "Gene Law" that it has been debating for years. In effect, the Gene Law has become the rope in a terrific tug-of-war, with the Green Party pulling for more environmental safeguards on one side and German industry pulling for freedom to pursue biotech on the other. Both the Greens and industry have considerable political clout in West Germany, so it's anyone's guess how, once it becomes the law, the Gene Law will actually work in practice. (Remind you of the situation at the EPA with TOSCA? See page 126.)

So what do the 100 German biotech companies do in the meantime? Go elsewhere to pursue biotech, for one thing. One example: The German multinational Bayer is establishing a biotech group in...Japan, the land where anything goes indoors and nothing goes outdoors. Bayer also announced that it will locate another biotech facility in Berkeley, Calif. — one of a handful of U.S. towns that once banned biotech.

Imagine fleeing radical politics in Germany by moving to Berkeley, where, as U.S. corporate writer Paul Bendix jokes, "On a good day, you can even find a demonstration against the nuclear family."

Deliberate Release *Ist Verboten*

Germany has experienced one release of a genetically engineered microbe.

In 1987, Dr. Walter Klingmüller of Bayreuth University released in a Bavarian pea field a *Rhizobium leguminosarum* strain with a marker gene added to it. He had added the marker gene *in vivo,* literally "in body." Germany's regulatory "hook" that yanks altered organisms in for an official look was *in vitro,* "in glass," so Klingmüller didn't think his release had to be reported to the authorities. And he certainly never dreamed he'd set off the firestorm of protest that culminated in June 1987 with a special Parliamentary Commission of Inquiry report called "Prospects and Risks of Genetic Engineering." The most sweeping among 170 recommendations was a call for a five-year ban on deliberate releases.

Though never endorsed by the German government, the ban on releases is *de facto* in effect three years later, as nothing genetically engineered has gone outside in West Germany.

In early 1989, however, the Germans almost experienced their second deliberate release. The Central Commission for Biological Safety, known as the ZKBS, approved a request by the Max Planck Institute for Plant Breeding to release in Cologne 37,000 petunia plants altered by biotech to contain some new genes. But, due to extensive protests by groups like the Green Party and the *Büeger Beobachten Petunien,* "Citizens Observe Petunias," the release was postponed until 1990 at the earliest. Green Party molecular biologist Christa Knorr explained the activists' position to *New Scientist* in March 1989:

"The necessity for the deliberate release stems from political rather than scientific reasons. This experiment with nice garden flowers is meant to habituate the German public to future release experiments."

Scientists countered with their reasons: Petunias are genetic workhorses, model systems that scientists can tinker with easily as they adapt biotech tools for use on much tougher genetic systems such as maize, wheat or canola plants.*

Freedom Yes But Not for Microbes

What's behind the extreme West German reaction to biotech? What motivates groups like "The Angry Viruses," which attempted to destroy a biotech laboratory at the Technical University of Darmstadt in January 1989? One thing is history, specifically the haunting shroud of Nazi Germany that Germans want to put behind them but can't. Especially when some activists consider biotech an extension of eugenics. The *New Internationalist,* a U.K.-based publication, published an article by an American freelance journalist in April 1988 called "A Molecular Auschwitz." Right up front in the piece, American biochemist Dr. Edwin Chagaff of Columbia University Medical School says:

*Tobacco plants, *E. coli* bacteria and brewers yeast are other favorite genetic "guinea pigs" that molecular biologists can alter with ease.

The angst avalanche completely buried biotech throughout the 1980s in West Germany.

"What I see coming is a gigantic slaughterhouse, a molecular Auschwitz, in which valuable enzymes, hormones and so on will be extracted instead of gold teeth."

Clearly not a comforting image for anyone, and particularly disconcerting for Germans.

But perhaps the biggest German force for opposition to deliberate releases of genetically engineered microbes is *Die Grunen,* the West German Green Party. Since its origins in West Germany in 1979, the Greens have become a major force in West German politics. In fact, more than 3,000 Greens now fill law-making positions at federal, state and local levels in German government. Three components of the German Greens' strongly anti-biotech platform, according to their own materials, are:

1. Prohibit companies from utilizing genetic engineering methods in research and production;
2. Immediately ban public financing of genetic engineering research and its applications in all areas; and
3. Enact a worldwide moratorium on the release of genetically modified organisms through action by national governments.

Have the German Greens had any impact on other European countries' attitudes toward biotech's products outside in the environment? Absolutely. No better example can be found than France, the next stop on our world "release" tour.

The French Connection

The 1987 release that lit the fuse under the Green explosion against all releases in Germany — the marked *Rhizobium* put into a pea field in Bavaria — was part of a three-country study that included France and Great Britain. So French and British scientists also released marked *Rhizobia* when the German team did in early 1987. The difference between the results of this release in France and Great Britain demonstrates that public reaction to any release may not reflect the attitude within the country where the release occurs, but the actions of aggressive neighbors.

French scientist Noëlle Amarger released marked *Rhizobia* in a field of peas near Dijon on March 13, 1987. Before Amarger's experiment went outside, few in France were even aware of it. Then West German Greens got wind of it and made it a *cause célèbre.* Soon, everybody in France heard about it, as *European Biotechnology Newsletter* detailed (7/22/87):

"The French press has taken over and widely broadcast a debate on deliberate release of genetically modified organisms, and the general public has been made aware that Rhizobium might well rhyme with plutonium ..."

Suddenly, the French people heard lots about deliberate releases because German Green representatives to the European Parliament*

*Originally created in 1957, the European Community (EC) now includes 12 countries: Germany, France, Italy, United Kingdom, Spain, Belgium, Greece, Netherlands, Portugal, Denmark, Ireland and Luxembourg. The 17-member European Commission is the only EC institution capable of drafting laws. It submits proposals on policy to the 12-member Council of Ministers, which is the essential decision-making body of the EC. Finally, the European Parliament has 518 members elected from the 12 member countries; while it has in recent years enjoyed greater influence on EC legislation, the Parliament has little actual power and is mostly perceived by European voters as "a relatively obscure debating society," according to *The New York Times* (6/19/89).

were opposed to it and made their views loud and clear. Benedict Haerlin, a German Green and spokesman for the environmental group Rainbow, accused the European Commission of "complicity in dangerous experiments involving deliberate release in nature of genetically manipulated bacteria."

Because the European Community (EC) had partially funded the three-country release study, this parliamentary protest became instant news in France. Just as the U.S. government had been caught without a clue of what to do in response to the media mayhem touched off in 1982 by a request to test ice minus outside — see page 114 — the French government was nabbed without a notion by an actual test in 1987. Dr. Amarger had sought no formal approval by federal authorities, because at that time none was required. It's still true in France that only commercial products, not academic experiments, must get legal approval for release from the Commission du Génie Biomoléculaire (Commission for Biomolecular Engineering), the national body that approves releases of genetically engineered organisms.

Release — *Oui* or *Non?*

Before 1987, France had established a reputation for being eager to entertain proposals involving field trials of biotech products — and this reputation did not vanish because of the embarrassment at Dijon. Indeed, Karl Simpson, European editor of *Genetic Engineering News,* wrote in the October 1988 issue:

"Among the many industrial attractions of France is a tolerant regulatory regime that takes a pragmatic view of legislation. The public is supportive of biotechnology and radical groups are practically non-existent. This situation might tighten up with the approach to the single European Market planned for 1992, but a number of French industrialists have told ***GEN*** *that they have no real fears in this area."*

But while French officials were willing to say that they supported the development and testing of biotech products, not one genetically engineered microbe has been released into the French countryside since Dijon. Crop Genetics International ran smack into this reality when its request to test InCide in southern France, a veritable shoo-in according to virtually anyone you asked on either side of the Atlantic at the time, was rejected in the spring of 1988. See page 78.

Where do French authorities stand on releases now? They still back the notion, but they're in a wait-and-see posture. They're waiting to see whether the dream of '92 can possibly come true.

The Impossible Dream

"A day will come when you, France; you, Russia; you, Italy; you, Britain; and you, Germany — all of you, all nations of the Continent will merge tightly, without losing your identities and your remarkable originality, into some higher society and form a European fraternity."

Soviet leader Mikhail Gorbachev quoting the words of Victor Hugo in a November 1989 speech to the Council of Europe

By January 1, 1993, Europeans expect to awake in a borderless Europe — at least when it comes to dealing with the rest of the world. Whether they really will is far too big a fish to fry here; but whether they will when it comes to the regulation of releases of genetically engineered organisms is not.

Most observers agree there's little chance Europe will become aligned on the issues of deliberate release any time soon.

For starters, no two countries anywhere think exactly alike on anything — and speculating that the 12 diverse countries that comprise the EC will do so on releases in a few short years is wishful thinking.

Second, as U.K. science writer Bernard Dixon observed in *Biotechnology Insight* (11/88):

"In Europe, regulations about the release of genetically altered organisms are a dreadful hotch-potch, ranging from an apparently highly restrictive law in Denmark to virtual absence of controls in Italy. EC efforts toward uniformity are moving forward, but much more slowly than the science and the technology are developing."

Third, since the early 1980s, the EC has been debating a proposal on deliberate releases of genetically engineered organisms. But, like the Gene Law in Germany, the EC directive on release has proven controversial. In September 1989, the Council of

Ministers, the EC's top decisionmakers, failed to reach an agreement on the outdoor directive, mainly because no one can decide what say countries should have in other countries' release decisions. The ministers sent the directive back to committee for further debate.

There's no telling when the ministers will agree on the directive governing releases. But it is clear that the cultural roots of various countries' reactions to releases — and to biotech — are unique, not really negotiable and not likely to disappear in just a few years.

Finishing the story of that three-country *Rhizobia* release will prove this point.

O.K. in the U.K.

First, to recap: The release of marked *Rhizobia* in early 1987 touched off a series of protests from German Greens that led to a halt of all releases — and temporarily to biotech — in Germany; German Green reaction to the French release of the same *Rhizobia* put a pall on microbial releases in France. What happened when the *Rhizobia* were, as the third part of the experiment, released in Great Britain?

Neither a bang nor a whimper was heard when scientists at the Rothamsted Experimental Station in Harpenden, Hertfordshire, let the marked microbes loose in the English environment in May 1987.

Not only did opposition from Greens in the European Parliament have little impact on the British *Rhizobia* release, the British press paid only perfunctory attention to it. Why? First, because the Greens were still, in 1987, a political non-entity in England — a situation that was bound to change, given the Green storm that was building across Europe in the late 1980s. And second, because everyone was watching something else that made the *Rhizobia* release look like what, scientifically at least, it was: no big deal.

First Viral Release

What was everyone watching? In September 1986, researchers at Oxford University's Institute of Virology had released a genetically engineered baculovirus in what was not only the United Kingdom's

The first genetically engineered virus to go outdoors made its debut in Scotland.

first release of a biotech microbe, but also the world's first release of a biotech-generated virus. Baculoviruses are viruses that infect certain species of insects. They have been used for decades as biological pesticides and were heartily endorsed as promising alternatives to chemical pesticides in 1973 by a joint meeting of the World Health Organization and the Food and Agriculture Organization.

The Oxford team spliced a marker gene into the baculovirus and released it to study the effects of gene-splicing on viral survival outside. This was a first step toward engineering faster and more potent viral pesticides. The release was conducted in Sutherland in the northern hinterlands of Scotland, a place that's populated by many more lodgepole pine trees than humans. That made sense, as the baculovirus in question kills the pine moth that kills pine trees.

What makes this release so interesting is that the scientists who conducted it, led by Institute Director Dr. David Bishop, waited until the British government had worked out voluntary guidelines for releases, which weren't finalized until a few weeks after Bishop's release was approved in June 1986.

What's more, Bishop's team did not get deluged by ferreting media; instead, Bishop went to great lengths to take the story to the press, which, when it covered the story at all, generally gave Bishop's openness as much or more play than the release he was proposing. Bishop even alerted such likely opponents as Friends of the Earth, which was, as Bishop told an international audience in Cardiff, Wales, in April 1988, "not overly supportive of this type of work, but...not antagonistic to it."

Here again, the only real protest came not from inside the United Kingdom but from outside — this time from the United States, in the person of perennial biotech protester Jeremy Rifkin. However, Rifkin's role was not significant, mainly because the British press is not accustomed to a crusading individual defining the debate over an entire technology, as Rifkin had been able to do throughout the early 1980s in the United States. As Matt Ridley, then science editor for the London-based *The Economist,* said in July 1986, "Our system in England is not as vulnerable to one person manipulating it via the courts and the press. It's also not part of our culture really — a shrill, Rifkin-type doesn't usually appeal to Brits."

Bernard Dixon recounts a revealing incident shared with him by a radio producer trying to find guests for a broadcast on Bishop's viral releases in late 1988. Bishop's team had successfully conducted releases of genetically engineered viruses in 1987 and 1988 — also with little protest and mostly favorable press.

Dixon explained what happened to the perplexed producer in *BioTechnology* (11/89):

"A spokesman for the Friends of the Earth (FoE) in Glasgow had expressed horror at the very idea of viruses being disseminated amidst the conifer-skirted lochs and bens of his beloved homeland. But when the producer approached FoE's headquarters in London, the response was very different. 'Actually,' Spokesman Number Two said, 'we rather favor this sort of biological control. It's far better than splashing toxic chemicals around.'"

This reminds us of those truisms that are rules of the road on our brief world tour: It is as unreasonable to say that all environmentalists think alike as it is to assume that all Greens think alike or that either automatically rejects all applications of biotech. Such sweeping conclusions hold no water.

England is Going Green

Like most of Europe, the United Kingdom is going Green like gangbusters. In fact, in the European Parliamentary elections held in June 1989, Britain's Green Party — actually the oldest in Western Europe, established in 1973 — garnered a larger share of the popular vote than any other Green party has to date.

But the Green wave sweeping across the United Kingdom does not necessarily mean that an anti-biotech or an anti-release wave will follow. Especially if the British government continues to construct a tough system of review for releases in which the public can participate — a rarity in the world release arena.

An influential July 1989 report on U.K. release policy by the respected Royal Commission on Environmental Pollution not only called for complete public disclosure of information on all releases, but also insisted that the public be brought into the review process whenever possible. Such full disclosure makes many scientists and most companies anxious, but the commission listened to the demands from the U.K. Green lobby for public access to the decision-making process. The premier U.K. biotech publication, *Biotechnology Insight,* explained (7-8/89):

"The Commission believes that public concern must be allayed in this way if there is to be any progress in the deliberate release of genetically engineered micro-organisms and plants for beneficial purposes in agriculture, environmental protection, pest control and so forth."

As for public participation in the release review process, Dr. Bishop put it plainly to *BioTechnology* (9/89):

"I can see no reason why Friends of the Earth, for example, should not be represented on the Advisory Committee on Genetic Manipulation, and every reason why it ought to be there."

The difference in what West German Greens have meant for biotech in West Germany and what U.K. Greens have meant for biotech in the United Kingdom is telling. The first have frozen their country solid on biotech; the second have helped make the United Kingdom perhaps the world's most open system for reviewing environmental releases. Both countries' Greens are equally active on environ-

mental issues; they just have reacted differently to biotech's products going outside.

Italy is Into Release

The Italians see green in biotech — money, that is. If West Germany represents the "anti" end of the European biotech spectrum, then Italy represents the "pro" end. In fact, Italy scarcely regulates biotech or releases, so it's difficult to say how many releases of exactly which altered organisms have taken place in Italy. But releases certainly have occurred there. "In fact," as well-versed U.S. science writer Yvonne Baskin wrote in the California publication *The Sun* (2/2/89):

"Italy is vigorously encouraging foreign firms to field test recombinant creatures there, earning the nation a reputation as 'the testing ground of Europe.'"

Denmark is Not — Yet

The Danish government made plain in its 1986 Environment and Gene Technology Act:

"Deliberate release of genetically engineered organisms or cells, including deliberate release for the purpose of experiments, shall not be allowed."

The following year the Danish Environmental Ministry cautioned:

"Genetic engineering can change our relationship to and our perception of the natural world. We could start to use our world in a different way ... If we have to alter nature — and in a specific case to what extent — is a matter on which it is important that everyone develops a view."

Clearly, the views expressed in these two statements would not lead you to Denmark were you interested in releasing biotech microbes. But, just as the 1980s closed, Denmark environmental Minister Lond Dybkjaer approved the release of two genetically engineered plants in 1990 by the Copenhagen-based food company Danisko. The minister used a little-known and never-applied rider in the Gene Act which states he may give his approval "in special cases."

Because they will be the first releases in Denmark — a country that has a strong Green lobby — Danisko's releases indeed will be "special." After all, Denmark has been widely regarded and reported to be the European country with the strictest policy against environmental releases anywhere. Consequently, the first release there will receive a lot of attention from German Greens in the European Parliament, and from the Danish press — and perhaps from the European press as well.

If it's Release Day it Must be Belgium

Belgium has released more biotech products into the environment than any other country — including the United States — and by far the most per capita. Its prodigious pace putting things out the door has been largely due to one company, Plant Genetic Systems (PGS) in Ghent. In fact, PGS Managing Director Walter De Logi told *Seed World* (5/89) that his company's 1988 program for field trials of genetically engineered crops was larger than all similar trials in the United States *combined.* PGS is busily working on genetically engineered microbes, but as yet has not released any.

To say that PGS has cooperated with and has the favor of the Belgian government is an understatement. What's more interesting is that PGS has jointly tested biotech products with researchers from both the public and private sectors in the United States, France, Spain, Italy, Holland and the United Kingdom.

Suffice it to say, releasing the products of genetically engineered organisms has not yet been a major issue for the public, the press or the government in Belgium.

Final European Stop: The Netherlands

Scientists in the Netherlands were the first anywhere to release a genetically engineered microbe into the environment — but it wasn't officially sanctioned or prohibited at the time so hardly anyone knew about it. It was in late spring of 1985 — two years prior to the first release of Frostban in California that was widely heralded as the world's first release — when Dr. Hans van Veen of ITAL, a government research institute in Wageningen, released bacteria with marker genes added to them.

Not a word was heard about this first release for two reasons. First, as van Veen explained in 1988, his first release "was not considered a recombinant DNA organism at the time" by Dutch authorities.* The second reason was, he says, "We were ahead of the public interest."

However, in the summer of 1989, an outdoor plot of some genetically engineered potato plants at ITAL was ripped up by a group called *Het Ziedende Bintje,* which means "The Raging Potatoes" or "The Seething Spuds." The climate for releasing genetically engineered organisms in the Netherlands had changed dramatically. Why? In two words: Dutch Greens.

Like the rest of Europe, the Dutch likely will wait until January 1, 1993, before testing any genetically engineered microbes outside. Like the Germans, the French and others, the Dutch don't want to be seen as moving ahead with such visible experiments before the European Community has approved the release directive, delineating what role countries can play in other countries' release decisions.

All said and done, since it takes at least a few years after any microbe is tested outside for it to become a product used outside, don't look for any biotech microbe to actually make it to the European market before 1995.

Then again, the same can be said about Japan, the United States and most of the rest of the countries in the world. In fact, only one country has had the "Gall" to sell a live, biotech microbe for use in the environment.

What's Up Down Under?

In a word, lots. We'll focus here on the story of the development of "NoGall," because it reveals Australia's attitude on deliberate releases, and provides a fascinating mini-case study.

"NoGall" is the trademarked name for the first live, genetically engineered microbe sold anywhere in the world. Now being marketed to Australian farmers, it's a bacterium that prevents a kind of plant cancer by killing its pathogenic relative, *Agrobacterium tumefaciens* (AT).

*A year later a similar release was subject to newly formed Dutch regulations on release.

AT has long been known to cause cancerous tumors at the base or crown of many woody plants, such as peach and almond trees. It causes "crown gall disease" by first slipping into a plant cell and then inserting one of its own genes into the plant cell's genome. As soon as the bacterium's gene gets into the plant cell's genes, it causes the plant cell to produce a chemical that allows the AT bacterium to proliferate cancerously.* The tumors or galls at the base of the plant are visible evidence that the AT bacterium has done its dirty but ingenious deed of gene insertion.

Turns out that another bacterium, *Agrobacterium radiobacter* (AR) — a close relative of AT — naturally produces an antibiotic that kills AT and thus prevents trees from falling to crown gall disease. Indeed, AR strain 86 has been used since the 1970s in orchards around the world to do just that.

Then, in the early 1980s, trouble. By way of the bacterial bazaars, a gene deal was made that was good news for AT and bad news for anyone with an orchard. AR began passing its antibiotic gene to AT, which in turn made AT immune to the once-deadly antibiotic. This enabled AT to go on causing cancer even in AR's presence.

So growers had only one option: to spray a chemical solution, the toxic chemical ethylene bromide, on threatened trees to kill the newly resistant AT.

Then Dr. Alan Kerr, Scottish-born plant pathologist at the renowned Waite Agricultural Research Institute in Adelaide, South Australia, entered this story. A world expert on AT, Dr. Kerr used biotech to essentially castrate AR by snipping out the gene for plasmid transfer, thus making it unable to pass its genes to AT or anything else. That meant no more immunity for AT; AR was back in business saving orchards. In fact, Kerr's new AR strain K1026 became the first open-air release of a genetically engineered microbe approved by the Federal Recombinant DNA Monitoring Committee, then Australia's main overseer of deliberate releases.

*AT's secret weapon is a plasmid taxi that transports its takeover gene into the plant cell. This AT vector or "gene taxi" has become, since 1983 when biotech allowed scientists to snip the cancerous gene out of it and thus "disarm" it, one of the most potent methods used by scientists to ferry foreign genes into various crop plants.

The entire review process, both locally and federally, of Kerr's small-scale release took less than nine weeks. On June 16, 1987, Kerr dipped the roots of 90 almond tree seedlings in a milky liquid containing K1026 and planted them in outdoor pots containing virulent AT.

Graeme O'Neill, science and technology reporter for the *Adelaide Age,* described this historic event:

The first release of altered bacteria in Australia was a microbial picnic.

"The day was bright, the event was dull ...

"Australia's first genetically engineered organism had been officially released into the environment, without fuss, fanfare, lawyers, scrutineers or protesters — indeed, even the local media ignored the moment."

No Look at NoGall

After Kerr's experiment, Friends of the Earth's Ian Grayson told the *Adelaide Advertiser* (9/26/87):

"It seems incredible that this technology is being introduced with hardly a murmur while U.S. environmentalists consider it as dangerous as nuclear power."

As wildly different as the reactions to Frostban and K1026 were, both bacteria are well-known organisms with well-defined single-gene deletions. In short, very similar experiments in very different countries under very different cultural circumstances led to vastly different reactions.

And results.

Back to the NoGall story. A company called Bio-Care based in Woy Woy, New South Wales, dubbed Kerr's K1026 "NoGall." In early 1989, Bio-Care was given permission by the New South Wales government to sell it, with "no questions asked" about toxicological and safety data, as Bio-Care Director Gary Bullard told *New Scientist* (3/4/89). Now any Australian farmer who wants to stop AT and stop using ethylene bromide to protect his plants can spray live, genetically engineered bacteria on them instead.

Green Light for Biotech

Since the first tests of K1026 outdoors, the Australian government has replaced its Recombinant DNA Monitoring Committee with the Genetic Manipulation Advisory Committee (GMAC). While more stringent than its predecessor, GMAC's review is voluntary and carries no legal sting. Furthermore, GMAC does not allow for much, if any, public debate about proposed releases, which makes Australian environmentalists worry that their country's laws and attitudes on releases are too lax. As Bob Phelps, Genetic Engineering Campaign Officer at the Australian Conservation Foundation, wrote in mid-1989:

"With the imminent shift of emphasis from laboratory work to deliberate release in field trials and commercially, there should be a public inquiry into genetic engineering. Sufficient uncertainties exist to justify a moratorium on all live releases, pending the outcome of that review and until a uniform national system of notification, assessment and monitoring is worked out and implemented."

Does this lax attitude mean that Australia is far from becoming green or not very active environmentally? Not at all; Australia claims more than 1,100 environmental groups. In 1989, Australian Prime Minister Bob Hawke became the first leader anywhere to appoint a globe-traveling "ambassador for the environment." Further, on July 20, 1989, Hawke delivered a major address called "Our Country, Our Future," which he later called "the world's most comprehensive statement on the environment." Mark Lonsdale of the Tropical Ecosystem Research Center in Darwin admitted to *New Scientist* (10/7/89): "Most ecologists in the world would be envious of a statement like this."

Remember our rules of the road? The degree to which any country has "gone green," or the degree to which a country protects its environment, doesn't necessarily predict its attitudes on the development of biotechnology for use in the environment.

What about countries where green doesn't mean anything yet? Countries like China and virtually the entire Eastern bloc, whose scientific expertise is difficult to ascertain and whose environmental concerns fall way behind a compelling desire to modernize industry and basically catch up with the 20th century. What do they think about biotech and deliberate releases?

Soviets' Open Invitation

The Soviet Union, under Mikhail Gorbachev, has restructured its biotech program and made it clear to the world that it is willing to test Western products of biotech in its environment for two reasons. First, to encourage Western expertise to come to the Soviet Union and help solve some of that country's many agricultural and environmental problems. And, second, to obtain whenever possible Western biotech training and technology.

The Soviet people see Western scientists as more trustworthy than their own. Or so Rod Greenshields, a British scientist who led a trip of British scientists interested in Soviet biotech to the Soviet Union in late 1989, averred, according to *New Scientist* (1/20/90). This is not to say that the Soviets will dump anything produced with biotech out the door just to see what happens, though some Western scientists who have visited the Soviet Union will say so privately.

However, when you have as many economic, agricultural and environmental problems facing you as Gorbachev does and you have political pressures to perform the likes of which the world may never have seen before, biotech looks a lot more like a potential problem-solver than the potential producer of environmental disasters.

The same is probably true of China, but factual information on Chinese biotech capabilities and attitudes on releases is tough to come by.

Where the Rest of the World Is

In our whirlwind world tour, we have touched on many of the most interesting spots in terms of releases. But there are many more — too many to cover in this *BriefBook*.

There are those countries with very sophisticated biotech capabilities that have already released biotech organisms — though few have released genetically engineered microbes — such as Canada, Mexico, Brazil, India, Argentina, Israel, South Africa and more. Some of these have had governmental debates reported by an aggressive press corps that doesn't miss much; others are more secretive or don't have a tradition of aggressive media — and some simply squash what the media can say.

Other countries have environmental attitudes and track records that don't lead one to believe that release debates will top their agendas if and when some scientist is ready to go outside with a biotech product. Some countries in this category are most of the rest of the Eastern bloc, China and many other fast-developing countries in Asia and Latin America. However, our rules of the road remind us not to read too much into any country's past record when predicting responses to biotech and/or releases.

E magazine called the 1990s "The Environmental Decade" in its debut issue in January 1990. The 1990s will also be the decade of debates around the world about biotech and releases of its products.

A Taste of Things to Come

Many of those debates will arise as a result of situations similar to this true story.

A brilliant woman in Thailand, using biotech, isolates a Bt gene that kills Anopheles mosquitoes. Anopheles is Greek and means "useless," but devilish would be a more apt descriptor, because it carries disease-causing organisms that cause much suffering. More than half the world's population lives in malaria-infested areas, and 250 million people suffer from the fever, chills and death that malaria-ridden mosquitoes deliver.

That scientist goes to develop her biotech expertise at a prominent university in Belgium. The newly-formed international team manages to splice the killer Bt gene into a perfect delivery vehicle — blue-green algae or "cyanobacterium"— that naturally inhabits rice paddies. Turns out that mosquitoes love to live and, more importantly, breed in rice paddies and other wet environs.

The potential for this discovery is mind-boggling: a proven, safe and specific biological pesticide against the Anopheles mosquito. What's more, it would replace toxic chemicals used for mosquito control — including DDT, which has been banned for years in most Western countries, but is still used by the thousands of tons in many countries.

Next, experts from an enterprising private, non-governmental group in the United States, a Belgian biotech firm, a prominent U.S. foundation and a leading U.S. university help devise a plan with Thai authorities to pursue the development of this microbial pest-killing package for distribution in developing countries. A biotech deal — involving the Far East, Europe and the United States — made for the benefit of suffering people.

Sounds too good to be true, but will it work?

Nobody can say, because no one is willing to try the blue-green, Bt gene-packin' pesticide outside. Why? All the parties involved are afraid of protests and subsequent media mangling when they are accused of releasing a genetically engineered microbe in a developing country — Thailand — that doesn't yet have clear regulations on releases.

Countries like Thailand, the Soviet Union, India, China, Brazil and numerous others that have both the capability to produce biotech organisms intended for use outdoors and fewer domestic obstacles delaying their release will likely prove to be the most volatile players in the nascent world debate about biotech and the environment.

The tough question is: What say does any country have in any other country's decisions about what novel organisms to release into the environment? This is a question that will explode in importance and visibility as the global applications of biotech continue apace throughout the 1990s and beyond.

In the meantime, three fundamental elements will remain central features of release debates around the world for years to come.

The Process...

Everywhere one thing is true. Genetic engineering makes most people a bit uneasy, if not downright distraught. It's powerful and people the world over don't understand it. And that makes them nervous.

The Press...

Then, when word gets out that a "genetically engineered mutant" is about to be released, that river of angst — see page 55 — starts to rise. Protests begin and the press arrives to cover sexy science being sold to unsuspecting or afraid consumers. Bingo, a release debate arises.

And the People

Finally, as is the case in the United States, so it is across much of the world: Public perception will be the eventual arbiter of all releases — indeed, of biotech. As Abraham Lincoln sternly reminded Stephen Douglas in an 1858 debate: "With public sentiment nothing can fail; without it nothing can succeed." Adjusted a bit for our story, without public sentiment, forget releases; with it, forge on — but always with considerable care.

8 A Ripe Time in History

"Microbes which at present convert the nitrogen into proteins by which animals live, will be fostered and made to work under controlled conditions, just as yeast is now. New strains of microbes will be developed and made to do a great deal of our chemistry for us."
Sir Winston Churchill, 1932

"Now at last, as it has become apparent that the heedless and unrestrained use of chemicals is a greater menace to ourselves than to the targets, the river which is the science of biotic control flows again, fed by new streams of thought."
Rachel Carson, 1962

Scientists are often the first to see those places to which leaders must lead the masses.

Churchill sensed the excitement that scientists of his era felt about the future and described for the masses what microbes might do for mankind *inside.* Thirty years later, Carson saw a time when microbes could do positive things for mankind *outside.*

In *Silent Spring,* she asked:

"For the microbes include not only disease organisms but those that destroy waste matter, make soils fertile, and enter into countless biological processes like fermentation and nitrification. Why should they not also aid us in the control of insects?"

Yet, years later still, few leaders have sensed the excitement in her vision, or the ripeness in biology that led Stanford University's Paul Ehrlich to write in *BioScience* in December 1989:

"There is no question that biology will be the science for the next few decades, both because of the nature of the human predicament and because of the rapid advances being made in areas as disparate as ecology and molecular biology."

Two Kinds of Leadership

Leadership comes in two forms. One: The people demand that government officials pursue a certain course — witness the people pushing their governments toward democracy throughout Eastern Europe.

Two: Leaders define a goal and take their people toward it — witness Mikhail Gorbachev making possible all of what is taking place in Eastern Europe and the Soviet Union.

When it comes to science, what brings any scientific starride of tomorrow into the realm of today's technology is not scientists. It's leaders. A prime example of true leadership: President John F. Kennedy's decision in May 1961 to put mankind on the moon before the 1960s were spent. As George F. Will editorialized in "Who Will Lead the Noble Quest?" in *Newsweek* (6/23/86):

"Looking around for something to do to symbolize fulfillment of his promise to get America 'moving again,' President Kennedy could look out a second-story window of the White House and there it was: the moon. It was good of God, or the big bang, to assign to Earth a single, accessible moon. But such obvious and popular goals are gone."

Not necessarily. God, or the big bang, also assigned to the moon a single, accessible Earth.

What the World Wants

Twenty years after that first small step onto the moon, the most important result did not even involve the moon, as Dr. Lewis Thomas recalled in *The New York Times* (7/15/89):

"But the moment that really mattered came later, after the equipment had been set up for taking pictures afield. There, before our eyes, causing a quick drawing of breath the instant it appeared on television screens everywhere, was that photograph of the Earth...

"That photograph, all by itself, was the single most important event in the whole technical episode, and it hangs in the mind 20 years later, still exploding in meaning."

That image of Earth has come to symbolize, more than anything else, the finite, no-dress-rehearsal deal that living on Earth is. Taking care of our home, our environment — making sure that Earth's fragile gown made of and by microbes does not come unraveled — is the clarion call of our time. The 1990s have already been dubbed "The Environmental Decade," "Earth Decade," "The Green Decade," "The Critical Decade" — call it what you will, the fact is, the world is excited about getting our collective home cleaned up ASAP.

A ripe moment in history it is.

Biotech But a Part

Environmental protection and cleanup is the kind of goal that's really impossible to be against, as Sandoz Crop Protection Corp. President and CEO Dale A. Miller said in October 1989:

"The irony is that, in my company, and I think in all agrichemical companies, we want exactly what our detractors want. Who can argue against removing pesticide residues from our food? Not me.

"The difference, of course, is the manner in which we get to our end point and what happens in between."

Biotech is part of what happens in between; it is not an end point — can't be, it's not even an "it." It is a powerful set of tools that, depending on how they're applied, can be central to many solutions along the long, hard road toward answering that call for a clean global house.

Perhaps the most important point about biotech and the environment is: Biotech must not be oversold as an environmental healer and it must not be undersold as an environmental destroyer. It can be both.

One who sees this clearly is Dr. Peter Raven, director of the Missouri Botanical Garden and a world-respected voice for the preservation of genetic diversity and the unique habitats that sustain it. Raven supported early legal action in the United States to stall the release of Frostban until effective regulations for releasing genetically engineered organisms were in place. But he also recognizes the importance of embracing the positive uses of biotechnology, as he made clear to an audience of environmentalists in late 1988:

"... any environmentalist who does not realize that genetically engineered cotton that would contain in itself the properties by which it could resist the pests that attack it, and which would, therefore, make unnecessary the application of millions of pounds of insecticides annually, has not carefully thought about the situation."

Cotton cultivation demands more than half of all the pesticides applied worldwide.

Calling All Leaders ...

*"'I am an environmentalist,' candidate George Bush declared in 1988. The statement could prove to be as relevant for us as John F. Kennedy's defiant **'Ich bin ein Berliner!'** at the Berlin Wall was for an earlier generation."*

Author and U.S. Ambassador Richard Elliot Benedick

"Even if you're on the right track, you'll get run over if you just sit there."

Will Rogers

"Gee, I wonder if any leaders even know we exist."

This is where CSI's story of biotech, microbes and the environment ends. And the real story begins.

If biotech is ever to play a positive role in environments around the world then leaders must do something. After all, this *BriefBook* has clearly illustrated one thing: People will never *demand* that genetically engineered microbes be released into the environment. They might not oppose them; indeed, they have even endorsed some when they understood both the microbe and the alternative. But they certainly won't lead the charge for releasing them.

So what must true leaders do now?

1. Demand a clear understanding of what biotech can and cannot do to microbes intended for use in the environment;
2. Commit the resources necessary to determine the real risks of releasing altered microbes into the environment;
3. Set goals — in the context of which, biotech becomes one of many tools, rather than an end or an issue in itself; and
4. Communicate clear goals, making people understand not only where they're being led, but also why that place is worth getting excited about. In short, lead.

...The People Will Decide

In the end, people will decide the fate of biotech, microbes and the environment. And it's clear that, when you mix genetic engineering and microbes meant for use outdoors, the river of public angst runs deep and can rise very quickly. As EPA administrator and environmental leader William Reilly said in his first major address on biotechnology in October 1989:

"Human anxiety about exotic new technology, no matter how potentially beneficial, is a primal force that must be recognized and dealt with if the promise of technology is to be realized. The public cannot be expected to accept — let alone embrace — what it doesn't understand, unless others it trusts ***do*** *understand and are reassuring."*

Those "others" are called leaders.

Some rivers will rise and nothing can be done about them. Legendary American singer Hank Williams Sr. once reassured his fans on a promotional poster for a concert: "I'll be there God willin' and the creek don't rise." Williams died in a car wreck en route to that concert.

It's difficult to say what makes the river of angst rise. But it's easy to say what makes it ebb: Knowledge, open debate and bold leadership.

Appendix

What is Biotech?

Split the word and it's easy to understand. "Bio" stands for biology, the science of life that includes all living things. "Tech" stands for technology, the tools and techniques in that big red toolbox to the right. "Biotech" is simply *applying* those tools to living organisms to get them to do something you want them to do.

Still, people complain that biotechnology is tough to define. It is, because it is not an "it," it's more of a "they." Though the singular word "biotech" is imprecise and often confusing, it is now part, not just of the English language, but of most languages throughout the world. "It" is here to stay.

Everything in the biotech toolbox enables humans to alter Nature's genetic handiwork and take further advantage of the things microbes — and anything else with genes — do of their own accord, without prompting. Generally, the older tools, those on the bottom shelf of the toolbox, have been used by plant and animal breeders, vintners, brewers and bakers for centuries. They allow only *gross guidance* of natural processes because they work in the *macroworld* of whole organisms: plants, animals and microbial cultures. It's the new tools — those on the top shelf used by today's genetic engineers — that permit for the first time *detailed direction* of genetic changes because they work in the *microworld* of cells, genes and DNA.

The biotech toolbox's top shelf — developed after 1973 — is to the bottom shelf what a microscope is to bifocals. It opened up a whole new world for scientists: the microworld of genes. Two important qualities separate the shelves, top from bottom, in the big red box. First, the top shelf adds specificity and precision unimagined before their development. Second, and truly revolutionary, the new tools in the box extend scientists' reach past the limited gene pools shared by species to the almost infinite gene ocean in all living things.

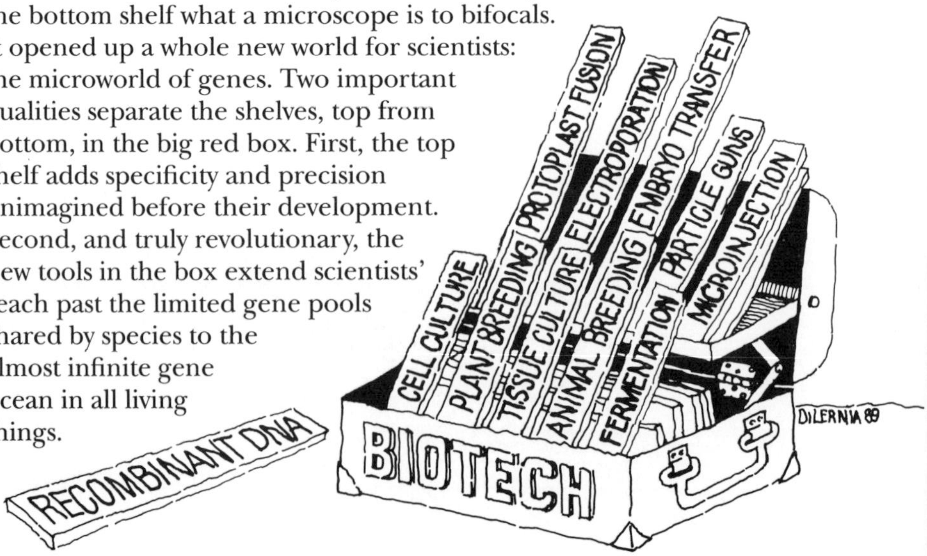

Agricultural scientists now have many biotech tools to choose from, both old and new.

One Giant Leap for Biology

When two California scientists spliced a gene from a toad into a bacterium back in 1973, they not only started the "biotech revolution," they catapulted biology over barriers Nature had been respecting for billions of years.

One crucial fact made this discovery possible: No matter where you find it, DNA is made from the same chemicals; it is the same in all living things. It is the expression of DNA — what it produces and when — that makes species different. Genes, themselves strands of DNA, are interchangeable because all living things understand the same basic DNA code. Whether you're a microbe, a plant, an animal or a human, your cells understand the same genetic language.

And, while rDNA is the media-darling quarterback on the top-shelf team of biotech tools, other players can bust through Nature's barriers. What links them all is their ability to move DNA from any living organism past natural barriers into the entire set of genes — called the "genome" — of any other organism. And they all play on the same field, namely the microscopic level of cells, those fundamental units of life that house genes and therefore DNA. A partial roster of the 21st century team includes:

Genetic engineers leap barriers that Nature erected billions of years ago, by moving genes between organisms that could never mate.

Microinjection: It's impossible to imagine what it's like to be a cell, but everyone knows what it's like to be injected with a hypodermic needle. Ouch! It's not pleasant, but it works. Well, scientists have devised needles so small that they can pierce individual cells and inject whatever they want into those cells. Including DNA from whatever source. It works too, but not all the time, because as small as cells are — remember, 1,000 bacterial cells strung end-to-end just reach across the period at the end of this sentence — even microscopic needle tips cause severe trauma, often killing the cell. This technique is used for animal, plant and microbial cells.

Electroporation: Everyone also knows what it's like to get a small jolt of electricity: really more frightening than painful. Well, scientists now use small surges of electric current to jolt cells that react, in a startled way, by opening small passageways in their outer walls for a split second. This allows whatever is in the immediate vicinity around the cell to slip inside, as clever gate-crashers do at a football game. Whoosh, they're in before you know it. As you might have guessed, the cell-crashers in electroporation are pieces of DNA, suspended in fluids.

Particle Guns: This one's almost unbelievable: Microscopic pellets — some made of gold and some of tungsten — are bathed in DNA and literally shot through cells. As these tiny bullets with DNA on board pass through cells, they leave some DNA behind. The aim is to get the DNA left behind by the bullet incorporated in the DNA of the cell, thus changing the genetic makeup of that cell.

Nothing is Out-of-Bounds

What all this means is that any gene from any organism can be placed into the genes of any other organism — *regardless of where it is perched in the evolutionary tree.* Whether taken from a complex organism in the topmost branches of the tree, or one from the thick lower boughs, or even one of the simplest root-dwelling microbes, it makes no difference. When it comes to moving genes from or to any part of the tree of life — regardless of which species, genus, family, order, class, division, or kingdom they normally come from — nothing is out-of-bounds for biotech.

This single fact about post-1973 or "new" biotechnology — its power to leap Nature's barriers — makes it, for many, frightening.

A Bacterial Bazaar

"We know some exquisite details about their [prokaryotes'] mechanisms of gene exchange in captivity, but do they really do these things in nature? They do, and the fascinating story is just beginning to unfold."
University of Iowa biologist Dr. Roger Milkman in *Science* (1/19/90)

When you enter the Bacterial Bazaar zone — microspaces where genes are traded, exchanged, stolen, lost and found in a series of surprising processes that have been going on for billions of years — you must remember two things.

First, scientists are just beginning to get a true image of what actually goes on between and among bacteria — and other microbes — in microspaces outside.

Second, if Bacterial Bazaars do exist, then there are billions and billions of them. So all of what follows takes place, if and when it does, in trillions of tiny places around the world.

Bazaar Behavior

Recall that bacteria or proks don't have nuclei — those central, contained places where most "living instructions" or genes reside in eukaryotic cells. What proks do have, indeed, what they are, is one long, circular strand of DNA inside a cell wall. Their genome — any organism's complete genetic recipe — is one molecule with only a cell wall between it and the world outside.

Unlike most euks, proks multiply by just splitting in half. So bacterial DNA is not handed down *vertically* to hybrid offspring from two parents; it's simply split in half along with the cell itself and thus passed one way in *horizontal* transfers. The result: two exact clones, not a parent and a child, but two identical twins. Under ideal laboratory conditions, bacteria can reproduce as often as every 20 minutes. In nature, where things are rarely ideal, bacteria can take hours, days, weeks — even years — to make more copies of themselves.

Still, when you add this unique ability to reproduce rapidly to the methods of gene exchange described in detail below, it's easy to see why bacteria have a more variable and surprising system of genetic exchange than do any other organisms.

Tricks in the Prok Trade

The main forms of gene transfer among bacteria are:

1. Transformation
2. Conjugation
3. Transfection
4. Lysogenization
5. Transduction

Let's quickly get to know each of these and their roles in the Bacterial Bazaar.

1. Transformation

When a bacterial cell dies or is caused by certain chemicals to "lyse" (pronounced "lice"), a gentle way of saying pop open, its one big strand of DNA can spill out in pieces. The lysed cell dies, but its DNA is now "on the market," so to speak, in the Bacterial Bazaar to be picked up by other still-intact bacteria. Most of the free-floating DNA gets destroyed outside cells, but some can float alongside a nearby bacterial cell and, with a quick exchange of some electrons — one of the forms of currency in this molecular market — pass into the neighboring cell.

Once inside the new host cell, the newly-acquired DNA fragments can wind up in the recipient's large circular strand of DNA — sometimes called the large replicon — and thus potentially transform the recipient bacterial cell's genome. That's transformation.

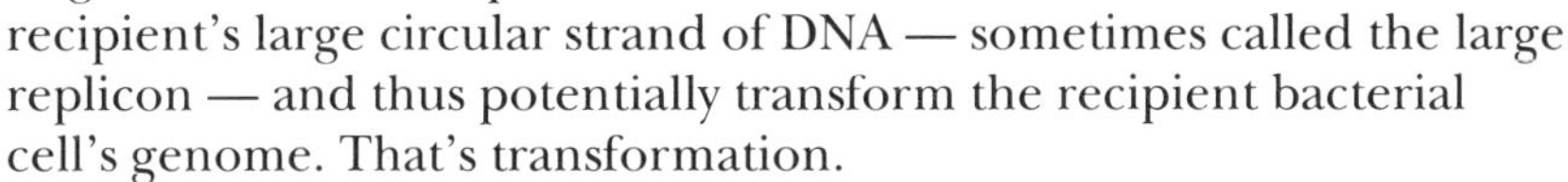

Whether the new DNA is advantageous or adaptive depends on what environment the acquiring cell finds itself in. If, for instance, a bacterium that lived in Yellowstone National Park were to obtain resistance to a popular pesticide that's never used in national parks, it certainly would not be an advantage. In fact, it might well be a disadvantage because, being so small and carrying relatively so little DNA and thus so few genes — 1,000 to 5,000 genes, it is estimated, while eukaryotes lug around 200,000 to 3 million — bacteria can't afford to carry "genetic baggage."

If, on the other hand, in the 122nd cell division, say, the bacteria with a pesticide resistance gene wound up in a place where pesticides were used, it would be pre-adapted for survival there.

2. Conjugation

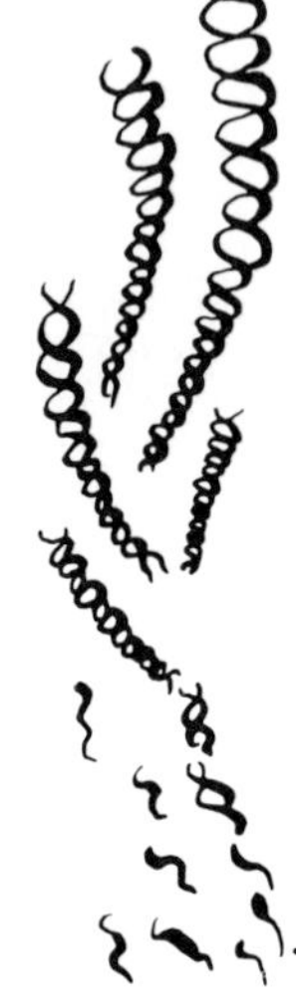

To understand the sexiest method proks use to make gene deals in Bacterial Bazaars, you must first get to know another Bazaar character: the plasmid.

Nothing more than small, circular rings of DNA that are separate from the main strand of DNA or large replicon — and thus called "small replicons" — plasmids are a kind of gene taxi that ferries genes about the Bazaar.

Conjugation is the one-way transference of DNA via plasmids through direct contact between two bacteria, usually of the same species. It's the closest that bacteria come to having sex and here's how "it" happens.

One type of plasmid is called "conjugative," and it controls the conjugating or the "getting together" of two cells to make a gene deal

up close and personal. Conjugative plasmids contain genes called the "sex factor" that can direct the cells they reside in to send out a "sex pilus" — a short tubelike structure — that literally touches another bacterial cell. Then, the first cell pulls its pilus back, drawing the second recipient cell closer and closer until they touch cell walls and, well . . . conjugate. Once together, so to speak, the donor cell pushes plasmids through its pilus into the recipient cell and "it" — gene exchange — happens.

Harley Schwadron, Reproduced by permission of Punch

Whew, for not actually having sex, proks certainly don't want for steamy situations.

The DNA that gets exchanged when conjugation occurs may either be plasmid DNA or DNA from the main host strand that the plasmid has "mobilized" — a nice way of saying popped in its back seat and kidnapped. And if the plasmid finds the climes favorable in the recipient cell's surrounds, it begins to make more copies of itself. When plasmids multiply this way in a given host cell, they turn that cell into another potential plasmid donor ready for the conjugation process. In this way, plasmids can be called "infectious" and resemble viruses — which we will meet shortly — as they move from cell to cell, wriggling through the hustle and bustle of the Bacterial Bazaar.

Adding to the shady side of the Bazaar, certain plasmids get around or conjugate a lot, not only with their own species of bacteria, but also with other species and even with bacteria from other genera — a surprising fact that's come to light in recent years. What do scientists call these social butterflies of the bacterial world? Promiscuous plasmids.

Finally, like viruses, plasmids are also known to confer such traits as resistance to chemicals, be they antibiotics or pesticides or natural toxins or herbicides. This ability to pass on to others such traits makes them very significant in the gene trades, and can be both good news and bad news to the scientists trying to exploit Nature's own gene taxis for various purposes. Plasmids and viruses are good news inside laboratories, where they already play central roles in the initial genetic engineering that moves genes of choice into or out of bacteria. But once altered bacteria are loosed outside, the same natural gene taxis can potentially take genes into other cells that neither Nature nor the scientists who engineered them intended them to.

3. Transfection

Now let's call out from behind the curtain and introduce the virus, one of Nature's craftiest characters.

Far smaller than even bacteria, viruses are measured in nanometers, 1,000 times smaller than micrometers used to measure bacteria and one million times smaller than millimeters, those littlest slashes between centimeters on the other edge of U.S. rulers. Viruses are efficient travelers: All they need to survive is just some DNA — some carry with them as few as 10 genes — and a protein coat to keep them together and that's about it.

Though small guests they may be, some viruses take over just about any cellular factory they wind up in. Once inside a cell, look out, "they're bad," to borrow a phrase from Michael Jackson.

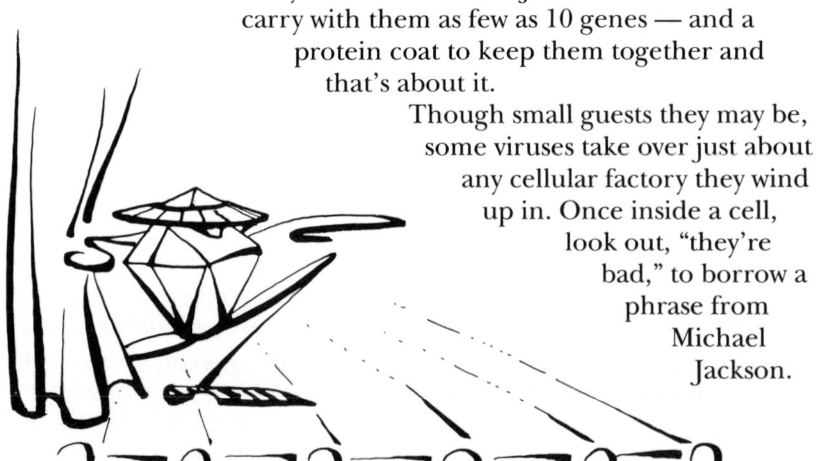

They whip off their coats, stick their own genes into the host cell's and take over the place. It's true: Specialized viruses insert their own genetic information into the host cell's DNA and, as a result, take over the function of that cell.

Viral takeover of a host cell's machinery has a name familiar to all: infection. Viruses that infect bacteria are termed "bacteriophages" — Greek for "bacteria eaters." They're called phages — pronounced like pages with an f — for short.

Now back to transfection. Since all cells can and often do house viruses, when a bacterial cell dies and releases its contents, viruses are often released too. And when one of these released viruses passes into another bacterial cell the process is exactly the same as transformation, but it's called "transfection." Why? Because what's being transferred in this special case of transformation is a virus that can *infect* the recipient cell.

How important are viruses in the movement of genes among organisms outdoors? No one can yet say. But check this: In August 1989, Norwegian scientists found a mind-boggling 10 million viruses in a single milliliter of lake water — that's *10 times* the number of bacteria found in the same sample and at least a staggering *1,000 times* more than had been found previously. This suggests that viruses are far and away the most abundant life form on Earth.

And transfection is just one way viruses, and the genes they ferry, move through a Bacterial Bazaar. Lysogenization is another and it's a more active process.

4. Lysogenization

Not all viruses are virulent; some are subtler guests. These friendlier forms of viruses are called "temperate phages." They too infect hosts and splice parts of their DNA into the host's main strand, but they don't necessarily kill it in the process. Once inside a host cell, temperate phages are called "prophages." Prophages can get carried along in duplicate cells when bacteria divide. Or, prophages can — after the bacteria carrying them lyse and loose their contents in a Bacterial Bazaar — pass via transfection directly into other bacterial cells.

Now, lysogenization begins when bacterial cells are stressed in some way — say a pesticide or herbicide is sprayed on them or maybe they just get hit with ultraviolet light from the sun. So stressed, those prophages that are already part of the host cell's central strand of DNA, start kickin' up their heels. Technically, they express what are called "conversion genes" that, as their name suggests, turn them from subtle guests into takeover artists. Now in charge, they direct the host cell to not only make many more copies of themselves (the prophages), but even to produce the protein coats that the newly created prophages will be dressed in to protect themselves as they move through the Bazaar.

What's more, and what is lysogenization, this tumultuous takeover and steady production of prophages continues until so many well-dressed new viruses are made that they literally lyse or burst the host cell and flood the markets, so to speak, throughout the Bazaar.

What's amazing is that these newly spewed viruses — called temperate phages because they are not part of a cell — can survive for long periods outside a cell, traveling freely on wind, wave, dust particle, pack mule or any other insect or animal. Wrapped in their protein coats, viruses are very resistant to environmental stress and can last a long time on their own.

Then, when they find a bacterial cell that has the right "receptor site," a chemically specific spot that viruses must find in order to infect cells, the process of infection can begin anew. A process that is, again, an act of gene transfer and recombination.

Quick quiz: Once inside their new host, these temperate phages are called what? Yes — prophages. As such, they can either direct the production of still more temperate phages; or they can, via the genes that they have inserted into the host cell's genome, confer upon it, through special proteins their genes code for, the ability to tolerate the stress — the application of pesticide or UV light, say — that caused them to start the active process of lysogenization in the first place.

Bazaar Break

Time out; let's make sure things are clear here by recapping viruses, the most resourceful, least understood characters in the Bacterial Bazaars.

Bacteriophages, or just "phages," are viruses that infect bacteria. As they cruise around Bazaars, they're called temperate phages. But, when they slip into a cell, they're called prophages and can either just hang out inside the cell or actually become inserted into the host cell's main strand of DNA.

It's these inserted prophages that, when stressed, activate their conversion genes and convert themselves into cellular party wreckers by forcing the host cell to produce more and more prophages. Finally, this continuous prophage production causes the host cell to pop, sending millions of new viruses into the Bazaar to slink around until they spy another unsuspecting cell to slip into and, when the time is right, take over.

In sum, phages are both good guys and bad guys in the Bacterial Bazaar: They help ferry genes around and get deals done, but they also can cut open — or lyse — bacteria whenever they express their conversion genes that convert them from subtler guests to "bad" cellular party wreckers. It all depends on what stresses they face.

Amazing place, this Bacterial Bazaar. And there's more.

5. Transduction

When lysogenization occurs and new prophages are created within a host cell, sometimes bits of the host cell's own DNA will get caught inside the protein coat that was created to encapsulate and protect the temperate phage when it moves through the outdoor Bazaar. Then, when the host cell literally bursts because of the proliferation of phages, those viruses with pieces of host DNA tucked into their coats get spewed into the Bazaar with the rest of the viruses. This process is called transduction. It means that sometimes when viruses move among bacteria, they take along non-viral DNA and thus further contribute to the constant genetic exchange going on throughout the Bazaar.

Finally, and most important, regardless of how viruses are moved, they are Nature's master genetic engineers: microscopic gene splicers that splice their own DNA into host cells' DNA, thereby producing recombinant DNA. And, since this natural genetic engineering has been going on in billions of Bazaars for billions of years, many scientists feel that introducing genetically engineered microbes into the environment is simply more of the same.

It may well be. But no one can say that based on evidence gained from man-made recombinants released into the environment because so few have been released. That's the catch: The only certain way to know how something will behave in the environment is to release it and find out.

A Bizarre and Sketchy Image

So there it is, a quick sketch of a fantastic Bazaar that scientists are only just now getting their first real glimpses of — and almost entirely because of the experiments that biotech allows them to conduct.

Exactly how and how frequently genes actually do move among bacteria — not to mention among and between bacteria and other organisms — is still a blurred portrait at best. Only time and study will illuminate all the amazing activity in microspace. But one thing is certain: Some activity exists — genes do move around. When they move, does it matter? Just how important are mobile genes in the greater scheme of things outdoors?

No one can say just now. But stay tuned to this fascinating and bizarre story.

Microbes' Role in Culinary History

"Civilization and the microbe go hand in hand."
Arthur Isaac Kendall

Truth is, civilization has always been married to the microbe, because being married to something means having it play a role, good or bad, in lots of things you do. And it's true that human experience with microbes began with a buzz, a baguette and some Brie.

This Bug's for You

No one can say for sure in which of the earliest civilizations that person woke up with the first and surely one of the nastiest hangovers ever. Whoever it was hadn't a clue that microbes — yeasts in this case — made the alcohol that made him or her feel so good and then so bad. But one thing is certain: the stumbling onto the effects of yeasts — namely, fermentation — was repeated in pounding head after pounding head around the globe, as the following examples illustrate:

6500 B.C. — Encrusted in the shards of a hunter-gatherer camp unearthed in 1983 by an Edinburgh archaeologist is a residue that could only be one thing — Neolithic heather beer. The recipe probably called for adding herbs to warm water and honey, then mixing these with ground barley and oats over a flame. The natural yeasts in the brew then converted the sugars from the cereal and honey into alcohol. According to Scottish legend, so highly prized was the secret recipe for heather ale that the last Pict clansman said to know it threw himself off a high cliff rather than surrender the secret to a rival band.

4000 B.C. — The Tigris-Euphrates cradle of civilization includes wine country, as viticulture has already been established. In Babylonia itself, beer later became the more popular drink, because the climate was better suited to the cultivation of grain than of grapes. It is certain that before 3000 B.C., malting and fermentation of grain was well-understood in Mesopotamia, where almost 40% of cereal production went into beer.

3000 B.C. — The Celts independently discover the art of brewing and Pliny the Elder, Roman author and natural historian, quips with his quill: "Western nations intoxicate themselves by means of moistened grain."

1750 B.C. — The oldest known recipe for beer is written on Sumerian clay tablets.

1300 B.C. — In a tomb in Xinyang, China, two bottles of wine are left behind by someone from the Shang Dynasty. Chinese archaeologists will find the wine in 1980.

700 B.C. — Early Mesopotamian descriptions of wine include "the divine liquid" and "unguent of the heart." Terms for beer were equally worshipful: "the plentiful," "the joy-bringer," "the heavenly," "the beautiful-good."

The role the discovery of microbial brews played in the civilizing of mankind remains unclear. Dr. Solomon H. Katz, professor of anthropology at the University of Pennsylvania, suggests that these discoveries led to the transformation from hunting and gathering societies to agricultural societies about 10,000 years ago. His theory can be inelegantly summed up: Wherever man first came home to a stiff drink he stayed. Perhaps. But whatever finally got humans to settle down, after they did microbes started to play even bigger roles in their lives.

Microbes Move into the Kitchen

Do you think that our ancestors knew the microbes involved in making the following food staples in various cultures?

- Cheese: *Penicillium roquefortii,* a mold, gives Roquefort cheese its pungent flavor. Legend has it that a peasant boy left a fresh piece of cheese from his lunch in a cave near the French village of Roquefort and when he came back a few weeks later, he found it moldy, but delicious. Another strain of the same genus, *P. camembertii,* gives Camembert its special qualities.
- Soy sauce: *Aspergillus sojae* and *A. oryzae,* also molds, yield — after a few days in a tank with roasted soybeans, wheat, a salty solution, and some bacteria and yeast — soy sauce. Tofu, tem-

peh and miso are made by similar processes using a variety of microbes.

- Yogurt: *Lactobacillus delbrueckii,* a bacterium, makes milk into yogurt. Before it became hip in Europe and the United States, yogurt had been a mainstay of diets in India and the Middle East for more than 1,800 years.
- Rice: *Anabaena azollae,* a kind of bacterium, forms a symbiotic relationship with the water fern Azolla. Before planting rice, farmers in China, India and Southeast Asia allow paddies to become covered with the fern. As the rice plants grow, they eventually crowd out and kill the Azolla, causing it to release its fixed nitrogen, which then fertilizes the rice plants.
- Vinegar: *Bacteria aceti,* commonly called the "mother of vinegar," causes the alcohol in alcoholic liquids, such as wine and hard cider, to turn to vinegar. The making of vinegar dates back as far as the making of booze. In the Book of Ruth in the Bible, Boaz invites Ruth: "At mealtime come thou hither, and eat of the bread, and dip thy morsel in the vinegar."
- Brewer's yeast: *Saccharomyces cerevisiae,* basic brewer's yeast, gives wine most of its flavors and many breads their height and stature.

It is certain that no one attributed all these pivotal products to their creators: microbes. How could they have? Microbes have only been visible since 1674 (see *Chronology* starting on page 187).

Mysterious Forces

"The problem of alcoholic fermentation, of the origin and nature of that mysterious and apparently spontaneous change which converted the insipid juice of the grape into stimulating wine, seems to have exerted a fascination over the minds of natural philosophers from the very earliest times."
Arthur Harden, *Alcoholic Fermentation,* 1923

The funny thing about the microbe/mankind marriage is that mankind hasn't known much at all about its microbial mates until very recently. Ancient peoples unwittingly celebrated the importance of microbes by creating myths, spells and gods to "explain" their invisible actions.

- Early Greeks gave credit for the invention of ale and wine to Bacchus, their god of wine.
- Egyptians — probably owing to the strong feelings they felt from its after-effects — attributed the art of alcoholic fermentation to Osiris, their god who was king and Judge of the Dead.
- Romans, who learned the art of brewing from Egyptians, gave ale the name "cerevisia," because it came from grain and grain was a gift from the goddess Ceres. This, by the way, is the origin of the modern Spanish word for beer, *cervesa.*
- According to Jewish tradition, the tree of life, planted in Eden by the evil spirit Sammael, was the grape vine, which was later saved from the flood by Noah. And it was Noah who taught the art of wine-making to the ancestors of the Hebrews. Genesis 9:20-21 says, "Noah was the first tiller of the soil. He planted a vineyard; and he drank of the wine, and he became drunk and lay uncovered in his tent."
- The Chinese recited spells during the most delicate part of the wine-making process — fermentation — to obtain supernatural aid for its successful completion.

Fermentation baffled brewsters and wigged out wine-makers because the microbes responsible could not be seen. "Fermentation" comes from the Latin *fervere,* to boil, which described the appearance of the bubbling yeast at the surface of a vat of wine or beer. The mysterious forces behind bubbling yeasts amazed alchemists of the Middle Ages, as Harden's text on fermentation testifies:

"Throughout the period of alchemy fermentation plays an important part; it is, in fact, scarcely too much to say that the language of the alchemists and many of their ideas were founded on the phenomenon of fermentation. The subtle change in properties permeating the whole mass of material, the frothing of the fermenting liquid, rendering evident the vigor of the action, seemed to them the very emblems of the mysterious process by which the long sought for philosopher's stone was to convert the baser metals into gold."

Indeed, microbial mysteries remained unsolved right into the 19th century. In 1804, one scientist wrote: "How staggered, even at this enlightened era, would some experienced brewers be, were they asked this one simple and plain question: 'In what does fermentation consist?'" And in 1837, the very year yeast was found to be a living organism, another scientist, W.H. Roberts, wrote:

"Discussion of the subject of fermentation would be of little real benefit to the operator; for confidently as many have asserted their knowledge of its secret causes and effects, the mystery in which its principles are involved continues to present an unpenetrated barrier; those who dogmatically profess to have encompassed this subtle and complicated subject only prove the extent of their ignorance and presumption."

These apparently mysterious forces would later be revealed as the work of microscopic toilers. And microbes affected other aspects of life — and death — as the *Chronology* starting on page 187 makes clear.

Chronology

1750 BC	Oldest known recipe — for beer — is recorded on Sumerian tablets.
600	Olive trees, along with unknown microbes, are brought to Italy by Greek settlers.
500	Chinese healers create the first antibiotic — moldy soybean curds — to treat boils.
250	Theophrastus writes of Greeks rotating their staple crops with broad beans to take advantage of enhanced soil fertility.
100 AD	Chinese growers use powdered chrysanthemum as the first insecticide.
594	An outbreak of bubonic plague, a bacterial disease later named the Black Plague or the Black Death, that began in 541 finally runs its course after claiming half of Europe's population.
1343	In the first act of biological warfare, Tartars catapult the bodies of Plague victims at invading Christians, hoping to spread the deadly disease.
1347	Fleas riding on black rats in the holds of Italian ships bring *Yersina pestis* (Yp), a bacterium, to the Sicilian port of Messina. Yp causes the Black Plague, which will kill 25 million Europeans — a third of the population — by 1351.
1520	Countless microbes move overseas on turkeys imported into Europe from America, on orange trees into Portugal from south China, and on maize (corn) into Spain from the West Indies.
1590	Dutch optician Zacharias Janssen invents the microscope.
1621	Potatoes from Peru are planted in Germany, another example of foreign microbes' finding new homes.
1665	Robert Hooke's *Micrographia* describes cells — viewed in sections of cork — for the first time. He named them cells because they looked like cells in monasteries.
1675	With a home-made microscope, Antonie van Leeuwenhoek discovers bacteria, which he calls "very little animalcules."
1790	United States passes first patent law.
1798	England's Edward Jenner coins the word "virus" to describe the matter that produces cowpox.
1802	German naturalist Gottfried Treviranus creates the term "biology."
1811	Organized bands of English handicraftsmen riot against the textile machinery displacing them, and the Luddite movement — led by a man they sometimes called King Ludd — begins near Nottingham, England.
1830	Scottish botanist Robert Brown discovers a small dark body in plant cells. He calls it the nucleus or "little nut."
1835	Charles Cagniard de Latour's work with microscopes shows that yeast is a mass of little cells that reproduce by budding. He thinks yeast are vegetables.

1840 The term "scientist" is added to the English language by William Whewell, Master of Trinity College, Cambridge.

1845 Late blight, a fungal disease afflicting potatoes, ravages Ireland's potato crop in 1845 and 1846; more than a million Irish die in the infamous potato famine.

1852 The United States imports sparrows from Germany as defense against caterpillars.

1857 Louis Pasteur begins the experiments that eventually prove definitively that yeast is alive.

1859 *On the Origin of Species,* Charles Darwin's landmark book, is published in London.

1862 The Organic Act establishes the U.S. Department of Agriculture (USDA) — formerly the Division of Agriculture in the Patent Office — and directs its commissioner "to collect ... new and valuable seeds and plants ... and to distribute them among agriculturalists." Untold numbers of foreign microbes begin entering the United States as a result.

1865 Augustinian Monk Gregor Mendel, the father of modern genetics, presents his laws of heredity to the Natural Science Society in Brunn, Austria. The scientific world, agog over Darwin's new theory of evolution, pays no attention to Mendel's discoveries.

1869 DNA is discovered in the sperm of trout from the Rhine River by Swiss chemist Frederick Miescher, but Miescher does not know its function.

Hemileia vastatrix, a microbial disease deadly to coffee trees, wipes out the coffee industry in the British colony of Ceylon (now Sri Lanka) and England becomes a nation of tea drinkers.

1877 German chemist Robert Koch develops a technique whereby bacteria can be stained and identified.

Louis Pasteur notes that some bacteria die when cultured with certain other bacteria, indicating that some bacteria give off substances that kill others; but it will not be until 1939 that René Jules Dubos first isolates antibiotics produced by bacteria.

1879 Chromosomes are discovered by German biologist Walter Flemming. Their function is not known.

1883 Francis Galton's *Enquiries into Human Faculty* introduces the term "eugenics" and suggests that humans can be improved by selective breeding.

The term "germplasm" is coined by German scientist August Weismann.

1884 Father Gregor Mendel dies after 41 years studying, with no scientific acclaim, the hereditary "factors" of pea plants; he said not long before his death, "My time will come."

1885 French chemist Pierre Berthelot suggests that some soil organisms may be able to "fix" atmospheric nitrogen.

1888 Dutch microbiologist Martinus Willem Beijerinck observes *Rhizobium leguminosarum* nodulating peas.

1889 The vedalia beetle — commonly known as the ladybug — is introduced from Australia to California to control cottonycushion scale, a pest that was ruining the state's citrus groves. This episode represents the first scientific use of biological control for pest management in North America.

1895 A German company, Hochst am Main, sells "Nitragin," the first commercially cultured *Rhizobia* isolated from root nodules.

1896 *Rhizobia* becomes commercially available in the United States.

1897 The Congressional Seed Distribution Program reaches its apex: 19 million packets of free seeds — with microbes on board — are franked in one year.

1900 The science of genetics is born when Mendel's work is rediscovered by three scientists — Hugo DeVries, Erich Von Tschermak and Carl Correns — each independently checking scientific literature for precedents to their own work.

1906 The term "genetics" is coined.

1909 Replacing Mendel's term "factors," geneticist Wilhelm Johannsen coins the terms "gene" to describe the carrier of heredity, "genotype" to describe the genetic constitution of an organism, and "phenotype" to describe the actual organism.

1910 Hucksters make a few bucks purveying inhalers touted to protect people from the poisonous cyanogen vapors emitted by Halley's Comet.

1912 American Chaim Weizman uses microbes to make the chemicals butanol and acetone, in the first application of microbial processes outside of the food industry.

1914 The first modern sewage plant, designed to treat sewage with bacteria, opens in Manchester, England.

1916 French-Canadian bacteriologist Félix-Hubert D'Hérelle discovers viruses that prey on bacteria and names them "bacteriophages" or "bacteria eaters."

1918 Worldwide Spanish flu epidemic kills 22 million — more than twice the number killed in World War I, the "war to end all wars."

1919 Hungarian Kark Ereky coins the term "biotechnology" to describe the interaction of biology with technology.

1923 More than 50,000 foreign plants have been introduced into the United States since 1862 by the USDA. Along with these plants came 90% of the pests that plague agriculture today; most are invisible microbes.

1925 Congress votes to cut off its expensive Seed Distribution Program, which had consumed more than 10% of the USDA's total budget in 1921, and the decades-old flood of free seed stops.

1926 Paul de Kruif's *The Microbe Hunters* becomes a popular book about bacteriology.

1928 Sir Alexander Fleming observes a culture of mold inhibiting growth of *Staphylococcus* bacteria in a petri dish. He names this mold "penicillin."

1933 American Wendell Stanley purifies a sample of tobacco mosaic virus (TMV) and finds crystals. This suggests, contrary to contemporary scientific opinion, that viruses are not just extremely small bacteria, for bacteria do not crystallize.

1938 The bacterium *Bacillus popilliae* (Bp) becomes the first microbial product registered by the U.S. government. It kills Japanese beetles.

1939 René Dubos, who will later enjoy international acclaim as an environmentalist, isolates gramicidin, an antibiotic, from a common soil microbe. His discovery helps cure a mastitis outbreak in the Borden Company's cow herd, including the famed Elsie, at the 1939 World's Fair.

The first large-scale deliberate release of bacteria into the environment takes place when Bp is sprayed over Connecticut, New York, New Jersey, Delaware and Maryland in an effort to arrest the damaging effects of the Japanese beetle.

1940 Howard Florey, Ernst Chain and others in England discover how to purify and preserve penicillin. The initial strains, which came from a moldy melon in Peoria, are submitted to X-ray radiation and ultraviolet light to produce more potent mutant strains.

American Oswald Avery precipitates a pure sample of what he calls the "transforming factor;" he has isolated pure DNA for the first time.

1941 Danish microbiologist A. Jost coins the term "genetic engineering" in a lecture on sexual reproduction in yeast at the Technical Institute in Lwow, Poland.

1942 President Franklin Roosevelt publicly denounces germ warfare as "an inhumane form of warfare." Privately, he approves a top-secret plan for the United States to develop biological warfare capability. A year later, the United States has a four-pound anthrax bomb.

1950 Aldrin, one of the deadliest chemicals available, is used by the U.S. government to attack the Japanese beetle in the midwest, replacing the bacterial insecticides that had been used earlier in the northeast.

The U.S. Army tests the spread and survival of "simulants," which are actually *Serratia marcescens* bacteria, by spraying them over San Francisco. Within days, one San Franciscan is dead and many others are ill with unusual *Serratia* infections, but the Army calls this "apparently coincidental." Similar tests are conducted in New York City's subway system, at Washington's National Airport and elsewhere.

1951 American Joshua Lederberg shows that some bacteria can conjugate or come together and exchange part of themselves with one another. He calls the material exchanged the "plasmid." He also discovers that viruses that attack bacteria can transmit genetic material from one bacterium to another.

1953 *Nature* publishes James Watson's and Francis Crick's 900-word manuscript describing the double helical structure of DNA, the discovery for which they will share a Nobel Prize in 1962.

1957 The Soviet Union launches Sputnik I, the first satellite.

1961 President John F. Kennedy decides to put mankind on the moon before the 1960s are spent.

1962 *Silent Spring,* a book by marine biologist Rachel Carson, galvanizes the first generation of environmentalists.

1963 E.P. Odin writes *Ecology,* the first textbook based on the principles of ecology.

1964 The International Rice Research Institute in the Philippines starts the Green Revolution with new strains of rice that double the yield of previous strains if given sufficient fertilizer.

1967 The National Academy of Sciences reports that the practice of adding antibiotics to animal food, while producing greater yields, may leave traces of antibiotics in meat, thus increasing drug resistance among bacteria.

1969 U.S. government takes steps to ban use of insecticide DDT.

On July 20th, the most famous descent down a ladder in history occurs when an American takes a small step onto the moon.

1970 The first Earth Day is celebrated on April 22 by 20 million people.

The Olin corporation stops manufacturing DDT.

1972 Paul Berg of Stanford University splices the DNA of viruses into the first recombinant molecules.

The U.N. Conference on the Human Environment in Stockholm thrusts the environmental movement into international view and, for a short time, draws worldwide attention to the urgent need to conserve the world's diminishing genetic resources, both plant and animal.

1973 The era of biotechnology begins when Stanley Cohen of Stanford University and Herbert Boyer of University of California at San Francisco successfully recombine ends of bacterial DNA after splicing a toad gene in between. They call their handiwork "recombinant DNA," but the press prefers to call it "genetic engineering."

Congress creates the U.S. Environmental Protection Agency (EPA).

1975 Scientists gather at Asilomar, Calif., for the first international conference on the potential dangers of recombinant DNA. They recommend that regulatory guidelines be placed on their work — an unprecedented act of self-regulation by scientists.

The number of new chemical compounds introduced to farmers each year peaks at 18.

1978 The first test-tube baby is born in the United Kingdom.

1979 On March 28, the nuclear reactor at Unit 2 of Three Mile Island undergoes a partial meltdown. Reporter Matthew L. Wald will write ten years later in *The New York Times* that Three Mile Island "was more than the breakdown of a nuclear reactor — it represented the failure of the entire system of nuclear regulation."

Asilomar restrictions are seen as too strict and are relaxed by the U.S. National Institutes of Health.

1980 In *Diamond v. Chakrabarty,* the U.S. Supreme Court upholds by five-to-four the patentability of genetically altered microorganisms, opening the door to greater patent protection for any modified life forms.

Biotechnology firm Genentech makes history on Wall Street when just 30 minutes after trading begins at $35 per share, the price per share hits $89 — the most meteoric rise ever. It closes at $71.25.

1981 Chinese scientists become the first to clone a fish — a golden carp.

1982 U.C. Berkeley's Steven Lindow is the first to ask permission to *deliberately release* genetically engineered microbes into the environment. He is immediately buried in an avalanche of angst — see page 51.

1983 Rita M. Lavelle, head of the Environmental Protection Agency's Superfund program to clean up toxic waste, is fired by President Ronald Reagan while under Congressional investigation for conflicts of interest and mismanagement. She later serves four months in prison for lying to Congress.

EPA Administrator Anne Burford is forced to resign.

NIH unanimously approves Lindow test.

1984 Federal District Court Judge John J. Sirica temporarily halts all federally funded experiments involving the deliberate release of recombinant DNA organisms.

EPA announces that if you genetically engineer any microbe intended for use outdoors, then you must pay a visit to EPA before you can legally test outside. For the complete story of biotech regulations in the United States, see *Regulating a Revolution,* starting on page 109.

Allan Wilson and Russell Higuchi of the University of California at

Berkeley become the first to clone genes from an extinct species. They clone genes from the preserved skin of a quagga, a form of zebra that has been extinct for a hundred years.

More than 3,000 people die after exposure to toxic fumes from a Union Carbide insecticide plant in Bhopal, India.

1985 An agricultural specialist with no experience in foreign affairs, Mikhail Gorbachev becomes Soviet leader after the death of Konstantin Chernenko.

Federal courts rule that private companies don't need National Institutes of Health's permission for field tests of genetically engineered organisms.

Mycogen scientists receive permission to test MVP, the first genetically engineered biopesticide to receive the Environmental Protection Agency stamp of approval.

1986 Researchers at the Institute of Virology at Oxford University release a genetically engineered baculovirus in what is not only the U.K.'s first release of a biotech microbe, but also the world's first release of a biotech-generated virus.

The USDA grants the Biologics Corporation the world's first license to market a living organism produced by genetic engineering: a virus used as a vaccine to prevent swine pseudorabies.

1987 Dr. Julie Lindemann sprays Frostban on an acre of strawberry plants in Brentwood, Calif., marking the first release of genetically altered bacteria in the United States. For a complete Frostban chronology, see page 52.

A West German Parliamentary Commission recommends a five-year moratorium on all releases of genetically altered organisms.

French scientist Noëlle Amarger releases marked *Rhizobia* in a pea field near Dijon; West German Greens get wind of it and make it a *cause célèbre.*

Earth First! vandalizes Brentwood strawberry patch, delaying AGS's second test of Frostban.

1988 Scientists from the U.S. Department of Agriculture and Crop Genetics International inject 2,200 corn plants with the world's first genetically engineered "plant vaccine."

Scientists at the University of Georgia's Savannah River Ecology Laboratory find *10 times* more genetic diversity than has ever been recorded in a single species of soil bacterium.

Deputy Health Minister Edwina Currie says that "most of the egg production" in Britain is infected with salmonella, a bacterium that can cause food poisoning. She loses her job and brings Britain's egg and poultry industry to the brink of collapse.

Canadian scientists discover that bacteria taken from the bodies of two explorers frozen in the Arctic since 1846 are resistant to modern antibiotics. This challenges current dogma that resistance to antibiotics is caused only by their overuse.

The U.S. Postal Service proposes to ban mailings of microbe samples capable of causing diseases. Of 90,000 samples mailed in 1987, only three leaks — of innocuous microbes — are reported.

A poll conducted by the Wirthlin Group reveals that 100 opinion leaders from across the United States believe biotechnology will be *the* technology of the 21st century.

1989 Bio-Care, a company based in Woy Woy, New South Wales, Australia, is given permission by the New South Wales government to sell "NoGall," and the world's first genetically engineered microbe designed for outdoor use goes on sale.

Exxon tanker Valdez strikes an Alaskan reef, causing one of history's worst oil spills.

The Administrative Supreme Court of the state of Hesse, West Germany, blocks the multinational chemical company Hoechst from completing a plant in Frankfurt for the production of human insulin because inside the plant technicians would be using genetic engineering.

A team of French and Israeli scientists use biotech to prove that plasmids can move freely even among different genera of bacteria. Thus, according to *BioTechnology* magazine, biotech may have "demolished" an accepted notion in microbiology — that plasmids move only among various species of bacteria.

A group called *"Het Ziedende Bintje,"* which means "The Raging Potatoes" or "The Seething Spuds," destroys an outdoor plot of genetically engineered potato plants at a government research station in the Netherlands.

A team of researchers at Lawrence Livermore National Laboratory in the United States produces the first direct images of DNA using a scanning tunneling microscope; the distance between successive coils in the helical ladder is about 1/5,500,000th of an inch.

1990 The second Earth Day is celebrated on April 22 by more than 225 million people.

The story of biotechnology, microbes and the environment enters its most interesting decade yet.

Glossary

aerobe A microorganism that grows in the presence of oxygen. A facultative aerobe can grow with or without oxygen, while an obligate aerobe must have oxygen to live. Compare anaerobe.

algae Eukaryotic microbes with plant-like characteristics. See *The Euks Are Coming* on page 20.

anaerobe An organism that grows in the absence of oxygen. Anaerobes cannot survive in the presence of oxygen.

antibiotic A chemical produced by one organism (typically a bacterium or fungus) that is harmful, if not fatal, to other organisms.

antimicrobial agent Any chemical or biologic agent that destroys or inhibits the growth of microorganisms.

bacillus A rod-shaped bacterium.

Bacillus thuringiensis (Bt) A bacterium that kills only organisms with alkaline stomachs, namely insects. With thousands of different strains, Bt are the workhorse bacteria in the microbial pesticide industry.

bacteriophage (also called **phage**) A virus that infects bacteria, sometimes causing their disintegration.

bacterium, pl. **bacteria** Any of a group of diverse and ubiquitous prokaryotic microorganisms. The simplest form of life, a bacterium is a single cell without a distinct nucleus.

Bacterial Bazaar A fictitious place where bacteria make gene deals of different kinds that involve various characters. See *A Bacterial Bazaar* starting on page 171.

bioaugmentation Increasing the activity of bacteria that break down pollutants by adding more of their kind. A technique used in bioremediation.

biodegradable Capable of being broken down by microorganisms.

bioenrichment Another bioremediation strategy that involves adding nutrients or oxygen, thereby bolstering the activity of microbes as they break down pollutants.

biologics Agents, such as vaccines, that confer immunity to diseases or harmful biological substances.

biomass Total dry weight of all organisms in a particular sample, population or area.

biome The largest ecological unit. A complex of terrestrial communities of very wide extent, characterized by its climate and soil.

bioremediation The use of microorganisms to remedy environmental problems. See **bioaugmentation** and **bioenrichment.**

biosphere The zone of air, land, and water at the surface of the earth that is occupied by organisms.

biostitute Adapted from "prostitute," slang for a biologist willing to say whatever a given constituency wants in order to forward a certain cause.

biota The animal, plant and microbial life defining a given region.

biotechnology See *What Is Biotech?* starting on page 168.

"bug" A slang term for microbe.

cell The basic unit of life; the smallest living structure capable of functioning independently.

chemophobia The fear of chemicals.

chromosomes Tightly coiled strands of genes (DNA) located in the nucleus of every eukaryotic cell. Each chromosome has a fixed number of genes, and every species has a characteristic number of chromosome pairs — humans have 23 pairs, mice have 19 and pea plants have 7.

clone An exact genetic replica, of one specific gene or of an entire organism.

cloning The process of asexually producing an identical copy of a parent organism, be it a cell or, say, a cutting from a tree. Bacteria, which reproduce by simply splitting in half, always produce clones.

codon A unit of three nucleotide bases that codes for one of the 20 amino acids. Strings of codons form genes and strings of genes form chromosomes.

colony A visible growth of microorganisms.

confinement Restricting the movement of most microbes during a small-scale release. Used instead of containment for field trials of microbes because it is impossible to totally contain microbes in an outdoor experiment.

conjugation Transfer of genetic material from one bacterium to another by means of close contact through a pilus, a tube-like structure.

containment The total control of a microbial application that is only possible in certain sophisticated indoor facilities. Compare **minimized dissemination.**

culture A particular strain or kind of organism growing in a laboratory medium. Also called a **cell culture.**

cytoplasm The living stuff inside all cells, but outside eukaryotic cells' nuclei.

deoxyribonucleic acid (DNA) The chemical chains that contain the genetic information in all organisms (with the exception of a small number of viruses in which the hereditary material is **ribonucleic acid — RNA**). The information coded by DNA determines the structure and function of an organism.

ecology (from Greek *oikos,* "house," and *logos,* "reason") The study of the interactions of organisms with their physical environment and with one another.

ecosystem A living system that includes the organisms of a natural community together with their environment.

ecotage Patterned after "sabotage," the destruction or vandalism of something the perpetrators believe will harm the environment.

endophyte An organism that lives inside another organism. An obligate endophyte must live inside another organism, whereas facultative endophytes can live inside or outside other organisms.

enzymes Proteins that control the various steps in all chemical reactions — usually speeding them along.

eukaryote or **eucaryote** A cell that possesses a definitive or true nucleus. All organisms except viruses, bacteria and blue-green algae are made up of eukaryotic cells. Compare **prokaryote.**

fermentation The controlled growth of microbes in open or closed systems for various purposes, such as the production of soy sauce, human growth hormone or beer.

fission An asexual process by which some microorganisms reproduce, involving the division of a single-celled individual into two new single-celled organisms of equal size.

fungicide An agent, such as a chemical, that kills fungi.

fungus, pl. **fungi** A microorganism that lacks chlorophyll. Examples of fungi include molds and yeasts. See *The Euks Are Coming* on page 20.

Gaia The Earth goddess of ancient Greece that British chemist Dr. James Lovelock uses to convey his hypothesis that the Earth itself is a living organism.

GEM Genetically engineered microorganism.

gene A segment of DNA specifying a unit of genetic information; an ordered sequence of nucleotide base pairs that produce a specific product or have an assigned function.

genetic marker A gene or group of genes used to tag microbes. For example, genes that code for resistance to antibiotics such as kanomycin or ampicillin are frequently inserted into plasmids used in gene transfer experiments to identify the transformed cells — when the antibiotic is added, those "tagged" cells are the only ones that remain alive.

genome The complete complement of genetic material in any individual.

genotype The particular set of genes present in an organism's cells; an organism's complete genetic recipe. Compare **phenotype.**

genus, pl. **genera** A group of very closely related species.

GEO Genetically engineered organism.

germplasm* The stuff of life: the material that controls heredity by governing the process of inheritance.

*For a complete description of the biological, historical, economic and political perspectives on germplasm, see CSI's *BriefBook:* Biotechnology and Genetic Diversity; ordering information is on the inside front cover of this *BriefBook*.

Green The color that now represents the exploding interest in the environment and is used to mean environmentally sound, environmentally conscious, and so on. Examples: green products, green consumerism, green labeling and greenwash — from whitewash, meaning false concern for the environment.

GMO Genetically modified organism.

host An organism harboring another organism.

hybrid The offspring of genetically dissimilar parents, whether cells, plants, animals or humans.

ice-minus bacterium The bacterium *Pseudomonas syringae* from which the gene that enables ice to form has been deleted. Plants sprayed with a suspension of ice-minus bacteria are able to withstand lower temperatures than they normally would.

in vitro Literally, "in glass." Describes biological reactions taking place in test tubes or any laboratory containers — all of which are artificial systems.

in vivo Literally, "in life." Describes biological reactions taking place inside living organisms.

microbe Any microscopic organism; a microorganism. From the Greek words *miko* for "small" and *bios* for "life."

microbial mats or **biofilms** Layered groups or communities of microbial populations.

microbiology The study of organisms of microscopic size (microorganisms).

micrometer The unit used for measuring microbes, equal to 10^{-6} m, which is one-millionth of a meter, one-thousandth of a millimeter. Abbr. **µm.**

minimized dissemination The EPA's phrase for "containment," which recognizes that released microbes can never be 100% contained.

mutagen An agent or process that can cause mutations, such as radiation or chemicals.

mutagenesis The process of mutation in the genetic material of an organism.

mutant A strain differing from its parent because of mutation.

mutation From the Latin word for "change," a genetic change that is caused by natural events or by mutagens. A stable change or mutation of a gene is passed on to offspring cells, whereas unstable mutations often kill the host.

mycorrhiza, pl. **mycorrhizae** Fungi that form symbiotic associations with the roots of higher plants.

nanometer A unit of length equal to one-billionth of a meter, or 10^{-9} m; 1 millimicrometer. Abbr. **nm.**

nitrogen fixation The transformation by certain symbiotic bacteria of atmospheric nitrogen into nitrogen compounds that plants can use as food. See **rhizobia.**

nodule Enlargement or swelling on the roots of legumes and other plants inhabited by symbiotic nitrogen-fixing bacteria.

nucleic acids There are two nucleic acids: DNA and RNA, made up of long chains of molecules called nucleotides.

nucleotides The building blocks of DNA and RNA. Each nucleotide molecule is composed of phosphate, sugar and one of four bases. These bases form codons, which when strung together form genes, which in turn link together to form chromosomes.

nucleus The central compartment in all eukaryotic cells that houses most of the DNA in higher organisms.

pathogen Any organism capable of producing disease.

phage See **bacteriophage.**

phenotype The observable characteristics of an organism. The phenotype results from the interaction between the genetic constitution (genotype) of an organism and its environment. Compare **genotype.**

plasmid Free-floating circular ring of DNA that can be integrated into genes and chromosomes. Plasmids can ferry genes among bacteria and thus are favorite tools of transformation for genetic engineers.

prokaryote or **procaryote** A bacterial cell lacking a true nucleus that usually has its DNA in one long strand. Compare **eukaryote.**

protozoa Eukaryotic microbes with animal characteristics. See *The Euks Are Coming* on page 20.

recombinant A cell or clone of cells resulting from recombination of genes.

recombinant DNA (Abbr. **rDNA**) As a product: Fragments of DNA from two different organisms, such as a human and a bacterium, spliced together in the laboratory into a single molecule. As a process: A broad range of techniques involving manipulation of the genetic material of organisms; "genetic engineering."

rhizobia Bacteria of the genus *Rhizobium,* which are commonly involved with leguminous plants in a symbiotic relationship that results in nitrogen fixation.

rhizosphere The soil region on and around plant roots — a zone of tremendous microbiological activity.

ribonucleic acid (Abbr. **RNA**) See **nucleic acids** and **deoxyribonucleic acid.**

sexual reproduction A process in which two cells — gametes — fuse to form one hybrid, fertilized cell.

smellfungus A word having nothing to do with science that means a critic or faultfinder.

species The taxonomic group of like individuals. In microbiology, variants within species are called strains.

spore A form of certain microbes that enables those microbes to exist in a dormant stage until conditions are right for them to "come back to life."

sterile Free of living organisms, namely microbes. Sterilization is the process of killing all life forms.

strain A different organism within the same species.

symbiosis The living together in close association of two or more dissimilar organisms; includes **parasitism** (in which the association is harmful to one of the organisms), **mutualism** (in which the association is advantageous to both), and **commensalism** (in which the association is advantageous to one and doesn't affect the other).

transduction, transfection, transformation See *A Bacterial Bazaar* starting on page 171.

transgenic An adjective describing an organism that contains genetic material from other organisms.

transposon A DNA sequence that is capable of transposition — the changing of location in a given genome.

vaccine A preparation of killed or debilitated microorganisms, or their components or products, that is used to induce immunity against a disease.

vector An agent, such as an insect, capable of mechanically or biologically transferring a pathogen from one organism to another. In genetics, any virus or plasmid into which a gene is spliced and subsequently transferred into a cell.

virulence The degree of pathogenicity exhibited by an organism.

virus A small genetic element, composed of DNA or RNA and protected by a protein coat, that is able to alternate between intracellular and extracellular states. Viruses do not have metabolic functions and cannot reproduce without first infecting and then taking over a living cell.

wild type Organism as found and existing in nature.

yeast A kind of fungi or microbe. See *The Euks Are Coming* on page 20.

References

*The books, reports, journals and magazine articles listed below will give you further information on the subjects and issues covered in this **BriefBook.***

Science

Note: Science books on microbes become dated quite fast. Thus it's safest to verify your information with active scientists. In biotechnology, nothing beats interviews and current science journals.

Brock, Thomas D. and Michael T. Madigan, Biology of Microorganisms, fifth edition, Englewood Cliffs, New Jersey: Prentice Hall, 1988. One of the top textbooks on microbiology.

Check, William, "The Reemergence of Microbial Ecology," *Mosaic,* Winter 1987/88. A fine and comprehensive overview.

Davison, John, "Plant Beneficial Bacteria," *Bio/Technology,* March 1988. A look at the good guys of the microbial world.

Hapgood, Fred, "Viruses Emerge as a New Key for Unlocking Life's Mysteries," *Smithsonian,* November 1987. A good, readable introduction to viruses, their discovery, and how they help us understand how cells work.

Lawrence, Eleanor, A Guide to Modern Biology: Genetics, Cells and Systems, Essex, England: Longman Scientific and Technical, 1989. An excellent everything-you-want-to-know book, but too deep for most nonscientists.

Margulis, Lynn and Dorion Sagan, Microcosmos: Four Billion Years of Evolution from Our Microbial Ancestors, New York: Summit Books, 1986. The major statement of the view that life has developed more through cooperation than competition, that bacteria are linked globally in a single gene pool.

Raven, Peter H., Ray F. Evert and Susan E. Eichhorn, Biology of Plants, fourth edition, New York: Worth Publishers, 1986. A can't-live-without textbook.

Regal, Philip J., et. al., Basic Research Needs in Microbial Ecology: For the Era of Genetic Engineering, Santa Barbara, California: FMN Publishing, 1989.

Sagan, Dorion and Lynn Margulis, Garden of Microbial Delights: A Practical Guide to the Subvisible World, Orlando: Harcourt Brace Jovanovich, Inc., 1988. A very enjoyable journey through microspaces — perfect for high-school science classes.

Sonea, Sorin and Maurice Panisset, A New Bacteriology, Boston: Jones and Bartlett, 1983. Similar to Microcosmos, but not as good nor as detailed.

Sonea, Sorin, "The Global Organism: A New View of Bacteria," *The Sciences,* July/August 1988. Cutting-edge thinking on microbes.

Risk Assessment and Introductions

Colwell, Robert, "Ecology and Biotechnology: Expectations and Outliers," in Joseph Fiksel and Vincent T. Covello, eds., Risk Analysis Approaches for

Environmental Releases of Genetically Engineered Organisms, NATO Advanced Research Science Institutes Series, Volume F, Berlin: Springer-Verlag, 1988. This article does a good job of laying out the different theories used to justify varying positions on the risk of rDNA releases.

Dixon, Bernard, Engineered Organisms in the Environment: Scientific Issues, York, Pennsylvania: The Maple Press Company, 1985. A lay summary of a 1985 symposium, this is still one of the best documents on the subject by one of the best writers in the business.

Douglas, Mary and Aaron Wildavsky, Risk and Culture: An Essay on the Selection of Technical and Environmental Dangers, Berkeley: University of California Press, 1982. The authors distinguish between how risk is perceived by two key segments of society — mainstream and fringe — but the text is heavy going.

Fleising, Usher, "Risk and Culture in Biotechnology," *Trends in Biotechnology,* March 1989. Fleising takes the arguments made in Douglas and Wildavsky and applies them to attitudes toward biotechnology.

Florman, Samuel C., Blaming Technology: The Irrational Search for Scapegoats, New York: St. Martin's Press, 1981. An interesting discussion of the fear of technology.

Krimsky, Sheldon, et. al., "Controlling Risk in Biotech," *Technology Review,* July 1989. An earlier version of this article appeared in *geneWATCH,* Vol. 5, No. 2-3. An article on the inadequacy of biotech regulations.

Levy, David J., "Technology, Politics, and the Responsibility of Intellectuals," *The World and I,* February 1987. Like the Florman book, this article takes a look at what we fear in the modern technological age.

Maranto, Gina, "Genetic Engineering: Hype, Hubris, and Haste," *Discover,* June 1986. Looks at the issues of risk and regulation and shows how little the debate has advanced in the last few years.

Mooney, Harold A. and James A. Drake, eds., Ecology of Biological Invasions of North America and Hawaii, New York: Springer-Verlag, 1986. A highly technical book on introductions, with the latest thinking on ecological theory. Especially useful to the general reader is the article by Philip J. Regal, which provides a humorous, but serious, look at the various arguments put forth about the safety of releases including the "pregnant pole vaulter" model.

Peterson, Cass, "How Much Risk Is Too Much," *Sierra,* May/June 1985. One of many good articles in the popular press that look at why people will accept some high risks and reject many low risks.

Sharples, Frances E., Evaluating the Effects of Introducing Novel Organisms into the Environment, prepared for United Nations Environment Programme (UNEP), 1986. Although slightly outdated, this is one of the most comprehensive surveys of information about deliberate and accidental introductions around. It also covers the thinking of a variety of ecologists about the problem of introductions in general and the assessment of environmental risks from the introduction of genetically engineered organisms.

Tiedje, James M., et. al., "The Planned Introduction of Genetically Engineered Organisms: Ecological Considerations and Recommendations," *Ecology*, April 1989. The best document to date outlining ecologists' approach to releases. Sober and realistic, it presents clear scientific arguments, minus the hype and polemic, about the releases issue. It outlines an approach for rating the relative risks of various kinds of releases that will help further the debate on regulation. Has an excellent bibliography.

Wells, H.G., The Food of the Gods, New York: Airmont Publishing Co., 1965. This is reading purely for fun. Wells's vision of what an environmental release could become.

Introduction of Recombinant DNA-Engineered Organisms into the Environment: Key Issues, prepared for the Council of the National Academy of Sciences (NAS), Washington, D.C.: National Academy Press, 1987. Known as the "Kelman Report" after the chairman (Dr. Arthur Kelman) of the committee that produced it.

Field Testing Genetically Modified Organisms: Framework for Decisions, Washington, D.C.: National Research Council, National Academy Press, 1989.

U.S. Regulation

Fanning, David W., Issues Raised by Biotechnology, Keystone, Colorado: Keystone Center, July 14, 1988. A good introduction to the issues surrounding regulation of releases. Has a good bibliography.

Hassebrook, Chuck and Gabriel Hegyes, Choices for the Heartland: Alternative Directions in Biotechnology and Implications for Family Farming, Rural Communities, and the Environment, Ames, Iowa: Iowa State University Research Foundation, 1989.

Mellon, Margaret, Biotechnology and the Environment, Washington, D.C.: National Wildlife Federation, 1988. The thinking on releases by one of the major environmental organizations. Mellon argues for extreme caution on releases and for a complete revamping of government regulatory programs.

Wildavsky, Aaron, Goldilocks is Wrong: In Regulation of Biotechnology Only the Extremes Can Be Correct, Berkeley: University of California Graduate School of Public Policy, September 1988.

Witt, Steven C., Regulatory Considerations: Genetically-Engineered Plants (Summary of a Workshop Held at Boyce Thompson Institute for Plant Research at Cornell University, October 19-21, 1987), San Francisco: Center for Science Information, 1988. An example of what a conference on risk can accomplish when goals are well-defined and scientists and regulators address specific issues and problems.

Keystone National Biotechnology Forum Interim Summary Report and Analysis of the Federal Framework for Regulating Planned Introductions of Engineered Organisms, Keystone, Colorado: Keystone Center, February 1989. The consensus document produced by dozens of professionals from all sectors examining the U.S. federal regulatory system throughout the mid-1980s. A must-have piece for those tracking biotech regulations.

International

Cohen, J.I., "Biotechnology Research for the Developing World," *Trends in Biotechnology,* Vol. 7, No. 11, pp. 1134-1139.

Dixon, Bernard, Engineered Organisms in the Environment, a lay summary based on the first international conference on the release of genetically engineered microorganisms (REGEM 1) held in Cardiff, UK, April 1988, available from the School of Agriculture, University of Aberdeen, UK.

Gibbs, Jeffrey, et. al., Biotechnology and the Environment: International Regulation, New York: Stockton Press, 1987.

Katz, J. Sylvan, Plant Biotechnology in Canada: Prospects for the 1990s, Saskatoon, Saskatchewan: Plant Biotechnology Institute, National Research Council Canada, 1989.

Organisation for Economic Cooperation and Development (OECD), Biotechnology: Economic and Wider Impacts, Paris: OECD Publications Service, 1989.

Organisation for Economic Cooperation and Development (OECD), Biotechnology and the Changing Role of Government, Paris: OECD Publications Service, 1988.

Swaminathan, M.S., Biotechnology and a Better Common Present, Kuala Lumpur: Asian and Pacific Development Centre, 1989.

Genetic Engineering and Biotechnology Monitor, an ongoing series of free publications compiled by the Industrial Technology Development Division, Department for Industrial Promotion, Consultations and Technology, UNIDO, Vienna, Austria.

History

Cooley, Arnold J. and J. C. Brough, Cooley's Cyclopaedia of Practical Receipts, Fourth Edition, Revised and Enlarged, London: John Churchill and Sons, 1864. Lots of fun for those who like looking at primary sources.

Dobell, Clifford, Antony van Leeuwenhoek and His 'Little Animals', New York: Russell & Russell, 1958. This is the book on Leeuwenhoek.

Forbes, R. J., "Chemical, Culinary, and Cosmetic Arts" in Charles Singer, E. J. Holmyard and A. R. Hall, eds., A History of Technology, Vol. I, Oxford: Clarendon Press, 1954.

Forbes, R. J., "Food and Drink," in Charles Singer, E.J. Holmyard, A.R. Hall, and Trevor I. Williams, eds., A History of Technology, Vol II, Oxford: Clarendon Press,

1956. These two articles have useful information on fermentation, especially brewing and winemaking, in ancient times.

Hare, Ronald, The Birth of Penicillin and the Disarming of Microbes, London: George Allen and Unwin Ltd., 1970. Must reading for those who have forgotten how much better life is since humans began to scientifically manipulate microbes.

Hughes, Sally Smith, The Virus: A History of the Concept, London: Heinemann Educational Books, 1977. An excellent source of easily readable information.

Judson, Horace Freeland, The Eighth Day of Creation: The Makers of the Revolution in Biology, New York: Simon and Schuster, 1979. Everything you ever wanted to know about the discovery of DNA and beyond. Long, but an engaging read.

Kendall, Arthur Isaac, Civilization and the Microbe, Boston: Houghton Mifflin Co., 1923. A prescient plea for a more balanced view of bacteria as essential to life on earth.

McGee, Harold, On Food and Cooking: The Science and Lore of the Kitchen, New York: Collier Books, 1984. An excellent, up-to-date source on the history of food, with lots of information on microbe-related processes.

Monckton, H. A., A History of English Ale and Beer, London: The Bodley Head, 1966. A good source of information on the history of English brewing.

Russell, E. J., Plant Nutrition and Crop Production, Chapter IV, "The Soil Microorganisms: Can They Be Controlled and Utilized?", Berkeley: University of California Press, 1926. Another fun book, it provides a look at the state-of-the-art of Rhizobia's use in the 1920s.

Miscellaneous

Anderson, Walter Truett, To Govern Evolution: Further Adventures of the Political Animal, Orlando: Harcourt Brace Jovanovich, Inc., 1987.

Carson, Rachel, Silent Spring, Boston: Houghton Mifflin Company, 1962.

Connor, Steve, "Genes on the Loose," *New Scientist,* May 26, 1988. Clear, concise article on some of the more prominent experiments underway (Bishop, Strobel, Wistar).

Finnegan, Jay, "All the President's Men," *Inc.,* February 1989. Excellent article on Crop Genetics International.

Hellemans, Alexander and Bryan Bunch, The Timetables of Science: A Chronology of the Most Important People and Events in the History of Science, New York: Simon and Schuster, 1988.

Myers, Norman, ed., GAIA: An Atlas of Planet Management, Garden City, New York: Anchor Books, Anchor Press/Doubleday & Company, Inc., 1984.

Nash, Roderick Frazier, The Rights of Nature: A History of Environmental Ethics, Madison: The University of Wisconsin Press, 1989.

"Biotechnology Survey," *The Economist,* April 30, 1988.

Recommended Periodicals

Biotechnology Insight, 12 Clarence Road, Kew, Surrey TW9 3NL, UK.

Biotechnology Notes, U.S. Department of Agriculture, OAB, Room 321-A, Administration Building, 14th and Independence Avenue, S.W., Washington, D.C. 20250.

Diversity, 727 8th Street, S.E., Washington, D.C. 20003.

geneWATCH, Council for Responsible Genetics, 186A South Street, Boston, MA 02111.

Genetic Engineering News, Mary Ann Liebert, Inc., 1651 Third Avenue, New York, NY 10128.

International Ag-Sieve, Rodale Institute, 222 Main Street, Emmaus, PA 18098.

New Scientist, 200 Meacham Avenue, Elmont, NY 11003.

Science, American Association for the Advancement of Science, 1333 H Street, N.W., Washington, D.C. 20005.

Science Digest, Family Media, 3 Park Avenue, New York, NY 10016.

Science News, Science Service, Inc., 1719 N Street, N.W., Washington, D.C. 20036.

The Sciences, The New York Academy of Sciences, 2 East 63rd Street, New York, NY 10021.

The Scientist, 3501 Market Street, Philadelphia, PA 19104.

South, South Publications Ltd., 13th Floor, New Zealand House, 80 Haymarket, London, SW1Y 4TS, England.

World Watch, Worldwatch Institute, 1776 Massachusetts Avenue, N.W., Washington, D.C. 20036.

Expert Sources

The individuals listed below constitute a treasure chest of valuable information, rich historical perspective, and seasoned opinion. You will find representatives from around the world in government, private industry, academia and the non-profit sector. Used well and often, this list of expert sources will add immeasurably to your understanding of the many issues and stories behind biotechnology, microbes and the environment.

Brian Ager
Director
Senior Advisory Group
on Biotechnology (SAGB)
c/o CEFCI
Avenue Louise 250, bte 71
B-1050 Brussels
Belgium
327-640-2095
Established in mid-1989, SAGB is, Ager explains, "the senior industrial forum for the development of European Community policy on biotechnology issues." Ager is a great place to start for European business sources in biotech.

Dr. Walter T. Anderson
Author and Environmentalist
1112 Curtis
Albany, CA 94706
(415) 526-5814
Anderson is the author of numerous books and articles on American politics and social movements. He is one of the few who can lucidly explain why environmental groups have such a difficult time forming policy on some biotech issues.

Philip S. Angell
Consultant
Browning-Ferris Industries
1150 Connecticut Avenue, N.W.
Suite 500
Washington, D.C. 20036
(202) 223-8151
A top aide to William D. Ruckelshaus for many years and a true student of public policy, Angell knows as well as anyone how Washington works — or doesn't work — on environmental issues.

Ken Ausubel
Executive Director
Bio-Remediation Services
621 Old Santa Fe Trail #10
Santa Fe, NM 87501
(505) 983-5549
A well-known filmmaker and authority on sustainable agriculture, Ausubel is developing bioremediation systems that involve many natural agents, including microbes.

Dr. Charles Benbrook
Executive Director
Board on Agriculture
National Research Council
2101 Constitution Avenue, N.W.
Washington, D.C. 20418
(202) 334-3062
A veteran of numerous NRC studies on all aspects of U.S. agriculture, Benbrook provides key insight into the influence such research has on the policy-making process.

Paul Bendix
Corporate Writer
261 Hamilton Avenue #304
Palo Alto, CA 94301
(415) 325-8847
Bendix is a communications consultant who mixes humor, intelligence and honesty with useful perspectives on biotechnology.

Dr. John E. Beringer
Professor
Department of Microbiology
University of Bristol
University Walk
Bristol BS8 1TD
United Kingdom
272-303030
When it comes to the deliberate release of GMOs in the U.K., no one is more knowledgeable about or involved in the issues than Beringer. He's an articulate "must" interview.

Dr. Sakarindr Bhumiratana
Deputy Director
National Center for Genetic Engineering and Biotechnology
Ministry of Science, Technology and Energy
Bangkok 10400
Thailand
662-2464988-9
If you have questions about biotechnology and deliberate releases in Thailand, NCGEB is the place to call.

Dr. David H.L. Bishop
Director
Institute of Virology and Environmental Microbiology
Mansfield Road
Oxford OX1 3SR
United Kingdom
0865-512-36
In September 1986, Bishop directed the world's first release of a genetically altered virus; thus he is a major player in release debates in the United Kingdom.

Dr. Guido Boeken
Public Affairs
Plant Genetic Systems
Jozef Plateaustraat 22
B900 Ghent
Belgium
91-35-8455
PGS is one of the most active biotech companies in the world in terms of releasing genetically engineered organisms. Boeken can lead you to numerous good sources inside PGS and throughout Europe.

Dr. Winston J. Brill
Winston J. Brill & Associates
4134 Cherokee Drive
Madison, WI 53711
(608) 231-6766
An outspoken scientist of international renown and an ardent biotech advocate, Brill is building a business around the process of scientific discovery.

R. Steven Brown
Director
Center for the Environment and Natural Resources
Council of State Governments
Iron Works Pike
P.O. Box 11910
Lexington, KY 40578-1910
(606) 252-2291
Tracking legislative trends at the state level is a major focus of the Council and tracking biotech is one of Brown's. He's always armed with facts and stats — a true find for the working journalist.

Steven Burke
Director
Education and Public Affairs
North Carolina Biotechnology Center
79 Alexander Drive - 4501 Building
P.O. Box 13547
Research Triangle Park, NC 27709
(919) 541-9366
An articulate product of the 1960s, Burke spins a steady stream of unexpected perspectives on biotech that make narrow-minded advocates on either side of any biotech issue re-examine their certainty.

Dr. Frederick H. Buttel
Professor of Rural Sociology
Cornell University
Warren Hall
Ithaca, NY 14853-7801
(607) 255-1676
One of the first to examine biotech's impacts on rural economies, Buttel has a growing family of former students now doing the same. Seeing biotech as neither panacea nor anathema, Buttel is a solid, informative source.

Mark F. Cantley
Director
CUBE Team
Commission of the European Community
Directorate General XII
Science, Research and Development
200 Rue de la Loi
B-1049 Brussels
Belgium
322-235-0749
Cantley follows research, events and trends in European biotech, which is like trying to keep track of a classroom of kids on a field trip to the zoo — chaotic, but always entertaining.

Dr. Ronald E. Cape
Chairman
Cetus Corporation
1400 53rd Street
Emeryville, CA 94608
(415) 420-3300
Few have Cape's credentials and credibility, and fewer still have his international perspective or his aggressive turn of phrase. One of the best interviews in all of biotech.

Dr. Peter S. Carlson
Vice President of Research and Development
Crop Genetics International
7170 Standard Drive
Hanover, MD 21076
(301) 621-2900
A brilliant scientist who can take any question and plant a pithy answer in real-world soil. An entertaining source with an impressive arsenal of analogies.

Dr. Jerry Caulder
President, Chairman and CEO
Mycogen Corporation
5451 Oberlin Drive
San Diego, CA 92121
(619) 453-8030
A farmer, former professional baseball player and respected scientist, Caulder left the comfortable climes of corporate America to start Mycogen. One of the most knowledgeable sources in biotech, he will gladly lunch on any sacred cow.

Dr. Joel I. Cohen
Genetics Biotechnology Specialist
ST/AGR/AP
Room 420E, SA18
Office of Agriculture
Agency for International Development
Washington, D.C. 20523
(202) 875-4219
Cohen can illuminate biotech's impacts on the Third World and steer you to innumerable sources overseas.

John J. Cohrssen
Senior Advisor to the Chairman
Council on Environmental Quality
Executive Office of the President
722 Jackson Place, N.W.
Washington, D.C. 20503
(202) 395-3742
Cohrssen is a little-known source of information about the history of U.S. efforts to throw regulatory bungie cords around the whale called biotech.

Dr. Robert Colwell
Professor of Biology
University of Connecticut
75 North Eagleville Road
Storrs, CT 06269-3042
(203) 486-4319
Skeptical and articulate, Colwell will find the ecological holes in your opinions on biotech, and, ever the professor, will teach you something as he does.

Dr. Peter R. Day
Director
Center for Agricultural Molecular Biology
Cook College, Rutgers University
New Brunswick, NJ 08903
(201) 932-8165
Day is British and knows much about British biotech and science policy. He also provides valuable comparisons of U.S. and European attitudes toward and capabilities in biotech.

Goffredo Del Bino
Head of Division for Chemicals and Biotechnology
Directorate General XI
Commission of the European Community
200 Rue de la Loi
1040 Brussels
Belgium
322-235-5443
EC efforts to develop policy on the deliberate release of altered organisms have been controversial to say the least, and no one knows more about them than Del Bino.

Dr. Alain Deshayes
Deputy Director for Plant Research
INRA
145 Rue de l'Université
CEDEX 07
75341 Paris
France
42-75-99-00
As the biosafety officer for INRA, Deshayes is involved with every deliberate release in France; thus, he's both a good source and a good link to other sources in the French government.

Kate Devine
Executive Director
Applied Biotreatment Association
P.O. Box 15307
Washington, D.C. 20003
(202) 546-2345
As the new leader of a new organization focused on bioremediation, Devine is on the cutting edge of biotech's applications in environmental cleanup.

Dr. Bernard Dixon
Science Writer and Consultant
130 Cornwall Road
Ruislip Manor
Middlesex HA4 6AW
United Kingdom
Ruislip (0895) 632390
Rarely do journalists have the depth of science background and expertise that Dixon has. On European biotech, Dixon is a solid source.

Jack Doyle
Director
Agriculture and Biotechnology Project
Friends of the Earth
218 D Street, S.E.
Washington, D.C. 20003
(202) 544-2600
Author of *Altered Harvest*, an analysis of biotech's impacts on agriculture, Doyle has distinguished himself as a leading critic of biotech's role in rural America.

Robert Ehart
Manager, Policies and Issues
Department of Public Affairs
CIBA-GEIGY Corp.
P.O. Box 18300
Greensboro, NC 27419
(919) 632-6000
A former state-level government professional turned corporate spokesman and reg-watcher, Ehart speaks in long, interesting paragraphs that prove he's weathered many a regulatory battle on state and national issues.

Dr. Ron J. Fenner
Director
Department of Research and Specialist Services
P.O. Box 8108, Causeway
Fifth Street Extension
Harare
Zimbabwe
263-4-704531
Few people know more about agriculture in Zimbabwe and southern Africa than Fenner; he explodes myths about developing-country agriculture and biotech's role in it.

Dr. Val L. Giddings
Senior Staff Geneticist
Biotechnology, Biologics and Environmental Protection
USDA/APHIS
6505 Belcrest Road, Room 850
Hyattsville, Maryland 20782
(301) 436-7602
Giddings can provide lucid detail on the genetics of fruit flies or behind-the-scenes political infighting on biotech regulations in D.C.

Dr. David Glass
Vice President
Government and Regulatory Affairs
BioTechnica Agriculture, Inc.
85 Bolton Street
Cambridge, MA 02140
(617) 864-0040
A veteran of release debates of the 1980s, Glass has shepherded field trials of genetically engineered bacteria and can tell you anything you want to know about what that involves.

Karl Glück
Gen-Ethisches Netzwerk
Winterfeldstrasse 3 D-1000
Berlin 30
West Germany
030-215-3492
Glück is the person to contact to learn how the Gen-Ethic Network is spearheading the coordination of European grassroots opposition to biotech.

Dr. Rebecca Goldburg
Staff Scientist
Environmental Defense Fund
257 Park Avenue South, 16th Floor
New York, NY 10010
(212) 505-2100
Here is a leading biotech-watcher who asks tough scientific questions of those who want to release organisms. Goldburg's non-shrill tone and solid science background make her especially effective.

Dr. Alan Goldhammer
Director, Technical Affairs
Industrial Biotechnology Association
1625 K Street, N.W., Suite 1100
Washington, D.C. 20006
(202) 857-0244
Always on the frontlines for the biotech industry, Goldhammer has a tough job and does it well. He's a very well-informed source.

Dr. Ralph W.F. Hardy
President
Boyce Thompson Institute
for Plant Research
Cornell University
Tower Road
Ithaca, NY 14853
(607) 257-2030
A senior player in both the biotech industry and in academic circles, Hardy has a powerful reputation that he mixes with a humble no-nonsense approach. He gets things done — in labs, in boardrooms and in debates.

Chuck Hassebrook
Program Leader
Stewardship, Technology and World Agriculture Program
Center for Rural Affairs
P.O. Box 405
Walthill, NE 68067
(402) 846-5428
A farmer and outspoken critic of biotech, Hassebrook speaks with hands-on authority about the role biotech products will and will not play on American farms.

Maureen Hinkle
Director of Agricultural Policy
National Audubon Society
801 Pennsylvania Avenue, S.E.
Washington, D.C. 20003
(202) 547-9009
Hinkle has lobbied on all aspects of U.S. agricultural policy for years, and she knows biotech well. What's more, she's one of the best, frank quotes in biotech.

Anne Hollander
Associate
The Conservation Foundation
1250 24th Street, N.W.
Washington, D.C. 20037
(202) 778-9702
For years a regulations watcher in Washington, Hollander can clearly explain the federal regulatory system to non-insiders.

Ernest T. Hubbard, Jr.
President
PureHarvest Corporation
P.O. Box 10287
Napa, CA 94581
(707) 224-6550
Hubbard has been an executive in two ag-biotech startups and now heads a company that will eventually put the products of biotech on dinner plates — using fewer chemicals. He's as interesting as the stories he tells.

Dr. Susanne L. Huttner
Associate Director
U.C. Systemwide Biotechnology
Research and Education Program
103 MBI
University of California
Los Angeles, CA 90024-1570
(213) 206-6814
Trying to increase public awareness and understanding of biotech in the largest agricultural state is a tall task that Huttner tackles on a daily basis. A solid source on public perceptions of biotech.

Dr. Calestous Juma
Director
Public Law Institute
P.O. Box 69313
Nairobi
Kenya
254-2-330098
Juma can clarify biotech's role in Africa — and he also brings expertise in global genetic resource debates to the discussion.

Dr. Kathleen Keeler
Associate Professor
University of Nebraska, Lincoln
School of Biological Sciences
348 Manter Hall
Lincoln, NE 68588-0118
(402) 472-2720
Though known more for her work on genetically engineered plants that might become weeds, Keeler is also expert on the environmental risks of GEOs.

Dr. Arthur Kelman
University Distinguished Scholar
Department of Plant Pathology
North Carolina State University
Box 7616
Raleigh, NC 27695
(919) 737-2711
A respected pathologist, Kelman is keenly aware of debates about the "new" biotechnologies. Unlike many scientists, he's not afraid to take positions.

Keystone Center
P.O. Box 606
Keystone, CO 80435
(303) 468-5822
Robert W. Craig, President
Michael T. Lesnick,
Senior Vice President
John R. Ehrmann,
Senior Vice President
The Keystone Center is *the* organization that has been bringing people from all perspectives together to discuss biotech since the mid-1980s. As such, Keystone's brass are ideal sources for insights into the perspectives and personalities of biotech players.

Dr. David T. Kingsbury
Professor
Department of Microbiology
School of Medicine and Health
Sciences
George Washington University
2300 Eye Street, N.W., Room 726
Washington, D.C. 20037
(202) 994-3484
The architect of President Ronald Reagan's policy on biotech, Kingsbury has been deeply involved in public debates about biotech policy and regulation. He remains one of the most articulate sources in biotech today.

Dr. Brian Kirsop
Executive Director
BioIndustry Association
One Queen Anne's Gate
London SW1H 9BT
United Kingdom
441-222-2809
One of numerous trade organizations that represent biotech firms in Europe, BA is a good place to obtain industry perspectives and contacts.

Dr. J.G. Kuenen
Department of Microbiology and Enzymology
Delft University of Technology
Juliana Laan 67
2628 BC Delft
The Netherlands
15-785308
Kuenen can explain the differences among various European countries' approaches to biotech and releases as well as he describes the contrasts between European and U.S. reactions.

Dr. Julianne Lindemann
Consultant in Agricultural Biotechnology Regulation
517 Everett Street
El Cerrito, CA 94530
(415) 526-4537
Few people have fixed a place in world history — especially while wearing a protective "moonsuit." Lindemann was the scientist-in-the-suit who sprayed Frostban outdoors in 1987 while the whole world — or at least much of its press corps — watched. She's shy, but full of knowledge.

Hannes Lorenzen
Agricultural Director
Les Verts
European Parliament
87-91 Rue Bellaird
BAP. ARD Bureau 319
B 1040 Brussels
Belgium
322-234-3408
Lorenzen coordinates policy on many biotech issues for the Green members of the European Parliament; so if you need a Parliamentary source to explain opposition to biotech, Lorenzen is one to call.

Dr. Bruce F. Mackler
General Counsel
Association of Biotechnology Companies
1120 Vermont Avenue, N.W.
Suite 601
Washington, D.C. 20005
(202) 842-2229
Founder of the biotech trade association that represents smaller biotech companies, Mackler is as fast as they come with a good quote and piercing point about biotech critics.

Sheldon Mains
State Planning Agency
Environmental Division
300 Centennial Building
658 Cedar Street
St. Paul, MN 55155
(612) 297-2376
A state government official who speaks his mind on biotech matters, Mains adds spice to any discussion of how states approach biotech legislation and regulation.

Dr. Lynn Margulis

University Distinguished Professor
Botany Department
University of Massachusetts
Room 322
Morrill Science Center
Amherst, MA 01003
(413) 545-3244

Margulis has built a very distinguished career constantly questioning accepted theories in microbiology. Indeed, she is a big reason that the explosive "Gaia" theory — that the earth itself is one big organism — is being reexamined by mainstream scientists.

Dr. William Marshall

President, Microbial Genetics Division
Pioneer Hi-Bred International, Inc.
4601 Westown Parkway, Suite 120
West Des Moines, IA 50265
(515) 270-3692

A leading scientist from a corporate giant on the cutting edge of biotech, Marshall asks tough questions about biotech, which makes him an unusually good industry source.

Terry Medley, J.D.

Director
Biotechnology, Biologics and Environmental Protection
USDA/APHIS
6505 Belcrest Road, Room 850
Hyattsville, Maryland 20782
(301) 436-7602

You cannot find a better source to explain what biotech regulators are up against. Medley can also provide an important international perspective to biotech regulatory discussions.

Dr. Margaret G. Mellon

Director
National Biotechnology Policy Center
National Wildlife Federation
1400 16th Street, N.W.
Washington, D.C. 20036
(202) 797-6891

Mellon has become perhaps the leading critic of U.S. attempts to regulate biotech. She has repeatedly called for a major overhaul of U.S. biotech regulations and thus is none too popular among those developing and implementing those regulations.

Dr. Elizabeth Milewski

Special Assistant to the Assistant Administrator
Office of Pesticides and Toxic Substances for Biotechnology
Environmental Protection Agency
401 M Street, S.W.
Mail Code TS-788
Washington, D.C. 20460
(202) 382-6900

Involved in the regulation of biotech since the late 1970s, Milewski can explain the EPA's inner workings and describe how the EPA fits into the history of biotech regulations in the United States.

Dr. John A. Moore

President
Institute for Evaluating Health Risks
100 Academy Drive
Irvine, CA 92715
(714) 725-9075

A key player in much behind-the-scenes battling on biotech regulation in the late 1980s, and former head of the EPA's pesticide program, Moore now spends his days immersed in the hottest science policy topic of the early 1990s: risk.

Robert B. Nicholas, Esq.
Partner
McDermott, Will and Emery
1850 K Street, N.W.
Washington, D.C. 20006
(202) 887-8000
With two decades of experience in Washington both on and off The Hill — much of that time working directly on biotech issues — Nicholas is a prime source for boiling the fat off Washingtonese and getting to the bones of policy and regulatory debates.

Dr. Thomas R. Odhiambo
Director
The International Centre of Insect Physiology and Ecology
P.O. Box 30772
Nairobi
Kenya
802501-3-9-10
One of Africa's leading scientists, Odhiambo is an expert source on biotechnology's potential role in Africa's many environmental problems.

Gilles Pelsy
Scientific Advisor to the Minister of Agriculture
75 Rue de Varenne
75700 Paris
France
45-55-95-50
Pelsy, like most French government officials, will not say anything revealing on the record. However, he has been deeply involved in the debate about the deliberate release of GMOs in France.

Bob Phelps
Genetic Engineering Campaign Officer
Australian Conservation Foundation
340 Gore Street
Fitzroy, Victoria 3065
Australia
613-416-1455
When it comes to environmental opposition to deliberate releases in Australia, Phelps can answer your questions and guide you to other key sources.

Dr. Mark Plotkin
Vice President for Plant Conservation
Conservation International
1015 18th Street, N.W.
Suite 1000
Washington, D.C. 20036
(202) 429-5660
Here is a scientist with extensive experience in Third World centers of genetic diversity. Plotkin's colorful stories and impeccable scientific credentials make him the perfect person to clarify biotech's role in international efforts to preserve diversity.

Dr. Philip J. Regal
Professor
Department of Ecology, Evolution and Behavior
University of Minnesota
109 Zoology Building
318 Church Street, S.E.
Minneapolis, MN 55455
(612) 625-4466
Regal is one of those scientists who can discuss the possible risks of releasing altered microbes in the same breath with Galileo's frictionless plane, but not lose you in the process. He's a well-known ecologist with a deep knowledge of biotech.

Jeremy Rifkin

President
Foundation on Economic Trends
1130 17th Street, N.W.
Suite 630
Washington, D.C. 20036
(202) 466-2823
Famed for his ability to impede the work of government agencies with lawsuits, Rifkin is the leading biotech critic in the United States. In early 1990, the *Utne Reader* described Rifkin as "the nation's most well-known Neo-Luddite."

Dr. Albert D. Rovira

Chief Research Scientist
CSIRO
Division of Soils
Private Bag No. 2
P.O. Glen Osmond
S.A. 5064 Australia
(08) 2749311
A good source for what's happening down under the ground in micro-spaces and for what's happening with biotech and microbes "down under" in Australia.

Dr. Peter M. Sandman

Director
Environmental Communication Research Program
122 Ryders Lane
Rutgers University
New Brunswick, NJ 08903
(201) 932-8795
One of the leading experts on risk, Sandman discusses it in a most entertaining, edifying fashion.

Dr. Setijati Sastrapradja

DIRBIONAS
Center for Research in Biotechnology
Jalan Raya Juanda 18
Bogor 16122
Indonesia
62-251-21038
"What," you wonder, "is going on in Indonesia with biotech in the environment?" Plenty, and "Seti," as her friends call her, will tell you all about it.

Dr. Robert E. Stevenson

Director
American Type Culture Collection
12301 Parklawn Drive
Rockville, MD 20852
(301) 231-5511
Who takes care of microbial samples that have been collected from all parts of the world? Stevenson is one such person, and is in touch with most of the others around the globe who do the same.

William "Skip" Stiles

Staff Director
Subcommittee on Department Operations, Research & Foreign Agriculture
House Agriculture Committee
U.S. Congress
1301 Longworth Building
Washington, D.C. 20515
(202) 225-0301
With 15 years on The Hill, a facile mind and a sharp wit, Stiles is one of the most knowledgeable sources on agricultural and biotech policy. Best of all, he puts Hill dribble into useful, funny context.

Dr. Trevor Suslow
Director of Microbial Pesticides
DNA Plant Technology Corporation
6701 San Pablo Avenue
Oakland, CA 94608
(415) 547-2395
Suslow is a scientist who chose the biotech business because he wants to change things — such as the amount of chemicals used in agriculture. Suslow weathered the Frostban wars of the mid-1980s and offers useful perspectives on them.

Dr. M.S. Swaminathan
President
IUCN
B4/142 Safdarjang Enclave
New Delhi 110029
India
044-455-339
One of a growing number of big-reputation scientists from the Third World, Swaminathan delivers serious statements with "Swami's," as he's known, inimitable soft touch.

Brian H. Sway
Executive Director
California Industrial Biotechnology Association
1112 I Street, Suite 380
Sacramento, CA 95814-2823
(916) 448-3862
An industry advocate in a state that boasts 40% of U.S. biotech companies, Sway is an excellent source on how California regulates biotech. He also knows the federal scene.

Dr. David Tepfer
Director of Research
Laboratoire de Biologie de la Rhizosphère
INRA CEDEX
78026 Versailles
France
331-30-21-38-35
It's difficult to find a scientist who is deeply involved in biotech, was raised in the United States but practices his trade in France, is articulate, and has a twin brother who is all of the aforementioned. Tepfer is one such rarity.

Fabio Teragni
Executive Director
Gruppo Di Attenzione Sulle Biotechnologie
Via C. Poerio 39
20129 Milano
Italy
022-940-6175
Many people say that little if any opposition to biotech exists in Italy. While Italy is certainly not as tough a place for biotech to flourish as, say, West Germany, Teragni, an antibiotech activist, is trying to change that.

Dr. James Tiedje
Professor
Department of Crop and Soil Sciences
Michigan State University
540 Plant and Soil Science Building
East Lansing, MI 48824-1325
(517) 353-9021
An expert on the ecological risks of releasing various organisms, Tiedje is an articulate scientist who can put risk discussions into lay terms.

Dr. Ken N. Timmis
Professor of Microbiology
GBF — National Research Center for Biotechnology
Mascheroder Weg 1
D-3300 Braunschweig
West Germany
495-31-61-81-400
Timmis helped form the European Environmental Research Organization to initiate more research — especially on environmental biotech — to combat global problems. If it involves bioremediation and is happening in Europe, Timmis can add rich perspective.

Dr. Sue Tolin
Professor
Department of Plant Pathology, Physiology & Weed Science
Virginia Polytechnic Institute and State University
Blacksburg, VA 24061-0330
(703) 231-5800
A scientist who has worked closely with the USDA on many biotech issues, Tolin provides unique insight — both as an insider and as an academic approaching what many academics would rather avoid altogether: regulation of science.

Nachama Wilker
Executive Director
Council for Responsible Genetics
186 South Street, Fourth Floor
Boston, MA 02111
(617) 423-0650
One of the best networkers in biotech activism, Wilker offers important perspectives on how U.S. and European opposition to biotech are similar and how they are different.